ECOtoxicology:
Ecological Dimensions

Chapman & Hall Ecotoxicology Series

Series Editors

Michael H. Depledge
Director and Professor of Ecotoxicology, Plymouth Environmental Research Centre, University of Plymouth, UK

Brenda Sanders
Associate Professor of Physiology, Molecular Ecology Institute, California State University, USA

In the last few years emphasis in the environmental sciences has shifted from direct toxic threats to humans, towards more general concerns regarding pollutant impacts on animals and plants, ecosystems and indeed on the whole biosphere. Such studies have led to the development of the scientific discipline of ecotoxicology. Throughout the world socio-political changes have resulted in increased expenditure on environmental matters. Consequently, ecotoxicological science has developed extremely rapidly, yielding new concepts and innovative techniques that have resulted in the identification of an enormous spectrum of potentially toxic agents. No single book or scientific journal has been able to keep pace with these developments.

This series of books provides detailed reviews of selected topics in ecotoxicology. Each book includes both factual information and discussions of the relevance and significance of the topic in the broader context of ecotoxicological science.

Already published

Animal Biomarkers as Pollution Indicators
David B. Peakall
Hardback (0 412 40200 9), 292 pages

Ecotoxicology in Theory and Practice
V. E. Forbes and T. L. Forbes
Hardback (0 412 43530 6), 262 pages

Interconnections Between Human and Ecosystem Health
Edited by Richard T. Di Giulio and Emily Monosson
Hardback (0 412 62400 1), 292 pages

ECOtoxicology: Ecological Dimensions

Edited by

Donald J. Baird
University of Stirling
Stirling
UK

Lorraine Maltby
The University of Sheffield
Sheffield
UK

Peter W. Greig-Smith
Ministry of Agriculture, Fisheries and Food
Lowestoft
UK

and

Peter E.T. Douben
Her Majesty's Inspectorate of Pollution
London
UK

Society of Environmental Toxicology and Chemistry

CHAPMAN & HALL
London · Weinheim · New York · Tokyo · Melbourne · Madras

Published by Chapman & Hall, 2–6 Boundary Row, London SE1 8HN

Chapman & Hall, 2–6 Boundary Row, London SE1 8HN, UK

Chapman & Hall GmbH, Pappelallee 3, 69469 Weinheim, Germany

Chapman & Hall USA, 115 Fifth Avenue, New York, NY 10003, USA

Chapman & Hall Japan, ITP-Japan, Kyowa Building, 3F, 2-2-1 Hirakawacho, Chiyoda-ku, Tokyo 102, Japan

Chapman & Hall Australia, 102 Dodds Street, South Melbourne, Victoria 3205, Australia

Chapman & Hall India, R. Seshadri, 32 Second Main Road, CIT East, Madras 600 035, India

First edition 1996

© 1996 Chapman & Hall

Softcover reprint of the hardcover 1st edition 1996

Typeset in 10/12pt Times by WestKey Ltd, Falmouth, Cornwall

ISBN-13: 978-0-412-75490-6   e-ISBN-13: 978-94-009-1541-1
DOI: 10.1007/978-94-009-1541-1

∞ Printed on permanent acid-free text paper, manufactured in accordance with ANSI/NISO Z39.48-1992 and ANSI/NISO Z39.48-1984 (Permanence of Paper).

Contents

Contributors

Donald J. Baird
Institute of Aquaculture
University of Stirling
Stirling FK9 4LA
UK

Peter Calow
Department of Animal and Plant
Sciences
The University of Sheffield
Sheffield S10 2TN
UK

Donald L. DeAngelis
National Biological Service
Department of Biology
University of Miami
PO Box 249018
Coral Gables
Florida 33124
USA

Michael H. Depledge
Department of Biological
Sciences
University of Plymouth
Drake Circus
Plymouth PL4 8AA
UK

Peter E.T. Douben
Royal Commission on
Environmental Pollution
Great Smith Street

London SW1 3B2
UK

Valery Forbes
Department of Life Sciences &
Chemistry
Roskilde University
PO Box 260
DK-4000 Roskilde
Denmark

Peter W. Greig-Smith
Ministry of Agriculture, Fisheries
and Food
Directorate of Fisheries Research
Fisheries Laboratory
Lowestoft
Suffolk NR33 0HT
UK

Paul Jepson
Department of Entomology
Oregon State University
2046 Cordley Hall
Corvallis
Oregon 97331-2907
USA

Peter Kareiva
Department of Zoology
University of Washington
Seattle
Washington 98195
USA

Lorraine Maltby
Department of Animal and Plant
Sciences
The University of Sheffield
Sheffield S10 2TN
UK

Tom Sherratt
Department of Biological Sciences
University of Durham
Sciences Laboratories
South Road
Durham DH1 3LE
UK

John Stark
Puyallup Research & Extension Center

Washington State University
Puyallup
Washington 98371
USA

Uno Wennergren
Department of Biology
Linkoping University
S-581 83 Linkoping
Sweden

John Wiens
Department of Biology
Colorado State University
Fort Collins
Colorado 80523
USA

Ecotoxicology is a relatively new scientific discipline. Indeed, it might be argued that it is only during the last 5–10 years that it has come to merit being regarded as a true science, rather than a collection of procedures for protecting the environment through management and monitoring of pollutant discharges into the environment. The term 'ecotoxicology' was first coined in the late sixties by Prof. Truhaut, a toxicologist who had the vision to recognize the importance of investigating the fate and effects of chemicals in ecosystems. At that time, ecotoxicology was considered a sub-discipline of medical toxicology. Subsequently, several attempts have been made to portray ecotoxicology in a more realistic light. Notably, both Moriarty (1988) and F. Ramade (1987) emphasized in their books the broad basis of ecotoxicology, encompassing chemical and radiation effects on all components of ecosystems. In doing so, they and others have shifted concern from direct chemical toxicity to humans, to the far more subtle effects that pollutant chemicals exert on natural biota. Such effects potentially threaten the existence of all life on earth.

Although I have identified the sixties as the era when ecotoxicology was first conceived as a coherent subject area, it is important to acknowledge that studies that would now be regarded as ecotoxicological are much older. Wherever people's ingenuity has led them to change the face of nature significantly, it has not escaped them that a number of biological consequences, often unfavourable, ensue. Early waste disposal and mining practices must have alerted the practitioners to effects that accumulated wastes have on local natural communities; for example, by rendering water supplies undrinkable or contaminating agricultural land with toxic mine tailings. As activities intensified with the progressive development of human civilizations, effects became even more marked, leading one early environmentalist, G. P. Marsh, to write in 1864: 'The ravages committed by Man subvert the relations and destroy the balance that nature had established'.

But what are the influences that have shaped the ecotoxicological studies of today? Stimulated by the explosion in popular environmentalism in the sixties, there followed in the seventies and eighties a tremendous increase in the creation of legislation directed at protecting the environment. Furthermore, political restructuring, especially in Europe, has led to the widespread implementation of this legislation. This currently involves enormous numbers of environmental managers, protection officers, technical staff and consultants. The ever-increasing use of new chemicals places further demands on government agencies and industries who are required by law to evaluate potential toxicity and likely environmental impacts. The environmental manager's problem is that he needs rapid answers to current questions concerning a very broad range of chemical effects and also information about how to control discharges, so that legislative targets for *in situ* chemical levels can be met. It is not surprising, therefore. that he may well feel frustrated by more research-based ecotoxicological scientists who constantly question the relevance and validity of current test procedures and the data they yield. On the other hand, research-based ecotoxicologists are often at a loss to understand why huge amounts of money and time are expended on conventional toxicity testing and monitoring programmes, which may satisfy legislative requirements, but apparently do little to protect ecosystems from long-term, insidious decline.

It is probably true to say that until recently ecotoxicology has been driven by the managerial and legislative requirements mentioned above. However, growing dissatisfaction with laboratory-based tests for the prediction of ecosystem effects has enlisted support for studying more fundamental aspects of ecotoxicology and the development of conceptual and theoretical frameworks.

Clearly, the best way ahead for ecotoxicological scientists is to make use of the strengths of our field. Few sciences have at their disposal such a well-integrated input of effort for people trained in ecology, biology, toxicology, chemistry, engineering, statistics, etc. Nor have many subjects such overwhelming support from the general public regarding our major goal: environmental protection. Equally important, the practical requirements of ecotoxicological managers are not inconsistent with the aims of more academically-orientated ecotoxicologists. For example, how better to validate and improve current test procedures than by conducting parallel basic research programmes *in situ* to see if controls on chemical discharges really do protect biotic communities?

More broadly, where are the major ecotoxicological challenges likely to occur in the future? The World Commission on Environment and Development estimates that the world population will increase from *c.* 5 billion at present to 8.2 billion by 2025. 90% of this growth will occur in developing countries in subtropical and tropical Africa, Latin America and Asia. The introduction of chemical wastes into the environment in these regions is likely to escalate dramatically, if not due to increased industrial output, then due to the use of pesticides and fertilizers in agriculture and the disposal of damaged, unwanted or obsolete consumer goods supplied from industrialized countries. It may be many years before resources become available to implement effective waste-recycling programmes in countries with poorly developed infrastructures, constantly threatened by natural disasters and poverty.

Furthermore, the fate, pathways and effects of chemicals in subtropical and tropical environments have barely begun to be addressed. Whether knowledge gained in temperate ecotoxicological studies is directly applicable in such regions remains to be seen.

The Chapman & Hall Ecotoxicology Series brings together expert opinion on the widest range of subjects within the field of ecotoxicology. The authors of the books have not only presented clear, authoritative accounts of their subject areas, but have also provided the reader with some insight into the relevance of their work in a broader perspective. The books are not intended to be comprehensive reviews, but rather accounts which contain the essential aspects of each topic for readers wanting a reliable introduction to a subject or an update in a specific field. Both conceptual and practical aspects are considered. The Series will be constantly added to and books revised to provide a truly contemporary view of ecotoxicology. I hope that the Series will prove valuable to students, academics, environmental managers, consultants, technicians, and others involved in ecotoxicological science throughout the world.

Michael Depledge
University of Plymouth, UK

1

Putting the 'ECO-' into ECOtoxicology

DONALD J. BAIRD, LORRAINE MALTBY, PETER
W. GREIG-SMITH AND PETER E. T. DOUBEN

This is an era of multidisciplinary science. One of the fastest growing young areas is ecotoxicology, which has developed mainly within the last quarter of a century by a synergy of classical toxicology and environmental chemistry. However, there are other traditions which can contribute insights to the ecotoxicologist's aims of understanding, predicting and managing the impacts of harmful chemicals on the environment. Not least, ecology should influence thinking within the subject, if it is to live up to its name.

So far, much of the development of ecotoxicology reflects ideas from the experimental tradition of toxicology and the dynamics of how chemicals behave in the environment. The emerging theories in the subject are thus rooted largely in the understanding of toxic mechanisms and demonstration of effects that occur in controlled conditions when all confounding factors are held constant. This is a sharp contrast to the context in which ecotoxicologists see the relevance of their science – to aid in the prediction of actual effects as they occur in the real world. At this level, it is the approach of ecology, with its emphasis on recognizing, observing and assessing the significance of changes in populations of plants and animals that may have most to offer.

We believe that the potential contribution of ecology to ecotoxicology has yet to be expressed. Ecotoxicologists have been slow to take full advantage of recent exciting progress in ecology, while those involved with the development of new ecological theories have often been reluctant to get involved in the applied problems of environmental contamination. In part, this reflects the regulatory procedures which attend much of the application of ecotoxicology

ECOtoxicology: Ecological Dimensions. Edited by D. J. Baird, L. Maltby, P. W. Greig-Smith and P. E. T. Douben. Published in 1996 by Chapman & Hall, London. ISBN hardback 0 412 75470 3 and paperback 0 412 75490 8.

and constrain the way in which new scientific ideas can be incorporated. The purpose of this book is to help redress the balance.

The six essays in the book were presented as invited lectures at a conference organized by the UK Branch of the Society of Environmental Toxicology and Chemistry–Europe (SETAC–Europe) in September 1994 at Sheffield, UK. The speakers were experts drawn from the two traditions of ecology and ecotoxicology, who were asked to look across the boundary between the subjects so as to shed new insights on the approaches taken by those on the 'other side'. We looked both for the injection of new theoretical concepts from ecology and for the emergence of opportunities to use ecotoxicological problems as a rich testing ground for ecological theory.

Of course, ecology is a very broad church, and this volume does not attempt to cover all areas where ecology and ecotoxicology overlap, for example, the fast growing area of behavioural ecotoxicology is not dealt with to any significant degree. Instead, we have selectively tackled a few of the 'big issues' which underlie the basic approaches at the ecological end of the ecotoxicology spectrum. These are all connected closely with the **variability** of ecological systems.

Variability pervades all natural systems and must be examined along a number of ecological dimensions. For ecotoxicologists, the relevant dimensions are: space and time, differences between species and differences between chemicals. In addition, we must recognize that ecotoxicology is both a science for the prediction of controllable risks and a means of retrospective interpretation of the effects of chemicals on ecological systems. These two purposes may require different approaches. Each of the chapters touches on the importance of some of these features.

Peter Calow (Chapter 2) sets the scene with a reminder that the measurements and observations made in ecotoxicological studies should be both reliable and relevant. This is more difficult than it might seem in systems that are complex and may be poorly understood. It is implausible to seek to model all effects and interactions; alternatively, he proposes that a series of practical 'rules of thumb' would allow the important ecological issues to be incorporated in guidance for the design of ecotoxicological studies. His tentative list of possible 'rules of thumb' is a challenge for both ecologists and ecotoxicologists to develop.

One of the most difficult aspects of ecotoxicological interpretation is to disentangle the effects of anthropogenic chemicals from 'natural' changes in ecological systems. This problem is discussed by John Wiens (Chapter 6), who examines the options for designing field studies that take account of the variability of environments in space and time. Importantly, he questions the notion that effects can be assessed simply by looking for change from a 'normal' condition since few systems are likely to be in a steady-state equilibrium. Different approaches may be needed to detect impacts on non-equilibrium systems. This chapter is an important cautionary tale for those involved in the design of ecotoxicological investigations.

At the heart of the need for an ecological input to ecotoxicology is the question "What constitutes a 'significant' effect?". This is crucial to the design and interpretation of studies, but has no simple answer. Peter Kareiva, John Stark and Uno Wennergren (Chapter 3) point out in their chapter that the traditional emphasis of toxicology on the deaths caused by chemicals may be inappropriate. In many cases, subtle sub-lethal effects that lead to impaired reproduction or delayed maturation can have important demographic effects in populations of both target and non-target species. They also illustrate that current developments in ecology, particularly the elaboration of 'stage'-structured (as opposed to age-structured) population models, should help to provide a framework for assessing toxic impacts and, equally important, the recovery of populations.

As well as a focus on population-level effects, ecology is concerned with changes to whole communities of animals and plants – well beyond the scope of the direct toxicological impact of chemicals on individuals. Some aspects of this are discussed by Kareiva, Stark and Wennergren, and by Donald DeAngelis (Chapter 4), who examines the complexities of the indirect knock-on effects that may spread throughout a community when one member suffers toxic damage. The problems of identifying all these consequences are formidable, but it is a subject of rapid progress among ecologists and may be one area where the needs of ecotoxicological problems could provide an opportunity for empirical research.

Even at the more familiar level of effects on local sub-populations of animals, ecotoxicologists do not currently have a firm basis for deciding what size of effect caused by a chemical 'matters'. There have been attempts to set thresholds for the amount of change (in numbers, or density, of animals) that are considered 'acceptable', but this has rarely been founded on strong ecological principles. Indeed, many ecotoxicological studies have failed to collect information in a way that would allow it to be entered into appropriate population models. The key to this problem is to take proper account of the scales of time and space. Paul Jepson and Tom Sherratt (Chapter 5) address the issue by reference to new developments in the theory of 'metapopulation' dynamics, showing its potential for the practical assessment of impacts of agrochemicals on non-target invertebrates. They show that by identifying the proportion of a habitat where animals are exposed to a toxicant, relative to their mobility and reproductive rate, it is possible to reduce the incidence of both false positive and false negative assessments of risk.

Impacts cannot be measured merely in terms of the numbers of animals or plants affected. Individuals vary in countless ways which may be relevant to the effects of chemicals, both by influencing the severity of an impact (e.g. only certain sections of a population may be vulnerable), and by changing the subsequent behaviour and performance of the population (because individuals are not removed at random). Valery Forbes and Michael Depledge (Chapter 7) discuss the implications of genetic and phenotypic variation within populations and suggest that variation in response to stresses such as toxicants may be just as important as the average impact. They also highlight an example in which the two traditions of ecology and ecotoxicology have evolved different meanings for the concept of 'sensitivity' – a good illustration of the benefits of a greater future dialogue.

The essays in this book provide many hints at the need for such a dialogue. As Kareiva, Stark and Wennergren point out, ecology has hitherto not been well-equipped to deal with the stochastic, non-linear, multidimensional aspects of its own subject. However, recent progress is providing the tools to do so. The time is now ripe to actively encourage us to put the 'ECO-' into ECOtoxicology.

2

Ecology in ecotoxicology: some possible 'rules of thumb'

Reliability and relevance are key criteria in making ecotoxicological observations and designing ecotoxicity tests. Reliability involves careful sampling design and requires detailed ecological knowledge of the subjects for sensible interpretation of results and for the long-term maintenance of cultures to support testing programmes. Relevance again involves a detailed understanding of the effect observed and its relationship to possible adverse impacts on the ecological systems of interest. In the latter case, it seems unlikely that it will be possible to consider detailed ecological models linking cause with effect for the risk assessment of the large numbers of chemicals that have to be scrutinized by ecotoxicity. Hence a series of general 'rules of thumb', guiding the design and interpretation of ecotoxicological observations and tests are suggested for further consideration and debate.

2.1 INTRODUCTION

Ecotoxicology is concerned with measuring and anticipating the effects of synthetic chemicals on ecological systems. It is an applied science in that it uses understanding from fundamental chemistry and ecology to assess the ecological risks posed by industrial processes, their products and byproducts. It can be used both retrospectively – are there effects and what is causing them? – and prospectively – are there likely to be significant effects?

Here I consider the ways that ecological science can be used to address these questions. In particular: to what extent can ecology be used in ecotoxicology to improve the reliability and the relevance of measurement and test systems? The magnitude of the challenge facing ecotoxicology in terms of the number

ECOtoxicology: Ecological Dimensions. Edited by D. J. Baird, L. Maltby, P. W. Greig-Smith and P. E. T. Douben. Published in 1996 by Chapman & Hall, London. ISBN hardback 0 412 75470 3 and paperback 0 412 75490 8.

of chemicals to be considered and controlled may mean that detailed ecological description and understanding is not always possible. Of more applicability may be rules that give guidance on criteria that are of general importance in nature. These are referred to as 'rules of thumb' and some likely candidates will be developed in the latter part of this chapter.

2.2 RELIABILITY

Reliability means that observations can be made in a controlled way and interpreted with confidence. For retrospective analyses this means the use of a combination of statistics and ecological understanding to design effective sampling programmes (e.g. [1]). Particular problems in programmes that compare sites separated in space and time are to distinguish significant variation that can be ascribed to pollutants from natural variation and to take appropriate account of the fact that the populations and communities in the separated sites might not have been similar irrespective of impact [1, 2]. For example, it is well known that the population densities of organisms vary systematically with season and microhabitat as a result of life-cycle events (e.g. [3]). Understanding this is crucial for distinguishing normal life-cycle and dispersional events from impact effects in comparisons of samples taken through the year and within particular microhabitats.

For prospective analyses, reliability involves the ability to provide test organisms in appropriate quantity and with appropriate quality on demand. This is a culture problem. It raises some non-trivial issues whose satisfactory resolution depends upon a sound understanding of the biology of the test organism; for example, in terms of its nutritional needs, optimum physico-chemical conditions for survival and reproduction, avoidance of disease, etc. (e.g. [4], and various reviews in [5]).

Reliability in the context of tests also implies that results can be repeated within laboratories and reproduced amongst them. That is to say, for the sake of both scientific and legal credibility, the same test when applied to the same substance should give similar results within laboratories at different times and when applied according to standard procedures in different laboratories. The variance within the system must therefore be identified and, apart from that due to operator fallibility, can usually be partitioned between that attributable to the intrinsic properties of the test system (genotype/species), that due to environment and that due to the interaction between these two. Identification and control of this variance can again involve a use of fundamental genetics and ecology [6, 7].

2.3 RELEVANCE

Relevance refers to the extent to which changes observed in systems under consideration – samples or tests – signal adverse changes in the ecological system of interest. Ideally, ecotoxicological programmes should start by defining what the system of interest is and what it is about that system that we are attempting to protect [7, 8]. But, more often, observations are made on convenient parameters and end-points. Thus, in retrospective studies, observations are most often made in terms of community structure parameters such as species lists, relative abundances and diversities because they are reasonably straightforward, and in prospective studies observations are most often made in terms of single species and using acute scenarios of high-dose exposure over short periods because these are relatively easy and inexpensive to effect [9]. Rarely are functional parameters measured in ecosystems, even though there is often a feeling that they may be at least as important as structural properties, and the accumulative effects of low doses on population variables, such as the timing and level of reproductive output, may be as important for population dynamics as survivorship impairment by high doses, especially in field situations.

Ecologically, the interest is in the state of systems consisting of collective groups of organisms, i.e. individuals in populations, and populations in communities and ecosystems. A reasonable ecological criterion of condition would therefore be that these systems should persist. Any parameter that provided an indicator of this would therefore be appropriate for ecotoxicological observation.

In theory this is straightforward and ecotoxicology should seek guidance from ecology on which of these indicators should be applicable (e.g. in terms of appropriate life-table variables [10]), but there are a number of complications. First, ecotoxicology, especially in prospective mode, is often not concerned with protecting particular systems. Much of the legislation that depends upon these kinds of tests is concerned with protecting ecological systems in general. So general measures of persistence are required. Second, though the criteria for population persistence are clearly defined by the theory of population dynamics [10], the same is not the case for communities and ecosystems. Here there is even debate about how structured and predictable these systems are [11]. So defining normal states in these cases is more problematic.

There is a tendency, therefore, in retrospective studies to measure a battery of criteria in a range of species and to compare between control and putatively impacted sites (e.g. the biological integrity criterion of Karr [12]). Yet this may waste time and resources; and still there is no guarantee that the most relevant (*sensu* above) criteria are being measured. Similarly, in prospective studies there are often attempts to use ever more sophisticated and sensitive end-points without these necessarily being relevant with respect to the persistence of populations, communities and ecosystems [7, 8].

Except in particular retrospective studies it seems unlikely that it will ever be possible to develop detailed descriptions and understanding of the ecological systems that are under threat. Given the enormous quantity of synthetic chemicals being used by society it also seems implausible that it will be possible to develop detailed models of the fates and effects of all existing and new substances so that detailed risk assessments (*sensu* [13]) can be carried out on all of them.

Instead, it is likely to be much more useful to use theory from ecology and other relevant areas of biology for guidance on the kinds of criteria that are likely to be generally relevant in defining harm to ecological systems and protecting them against pollution. This is especially so with respect to prospective tests concerned with screening chemicals for likely effects on the environment at large.

2.4 'RULES OF THUMB'

This general guidance in designing monitoring, surveillance and survey work, or designing and implementing prospective tests, might be conceptualized as a series of 'rules of thumb' – not necessarily applicable in all circumstances but likely to be applicable in most circumstances. Some candidates are summarized in Table 2.1. These are presented as questions because they are not intended to be definitive at this stage, only to suggest areas where 'rules of thumb' would be very helpful in the design, implementation and interpretation of ecotoxicological analysis. They deserve some justification and further comment.

Table 2.1 Some possible 'rules of thumb' that could be important for ecotoxicology

1   That structural properties of ecological systems are more important than functional ones

2   That the state of a few particular species in communities is likely to be more important than effects on a large number of species for community structure and/or function

3   That the direct effects of toxicants on survival and reproduction are more important than indirect action due to adjustments in predator and/or prey competitor–competitor interactions

4   That aquatic species differ in sensitivity but in general by no more than x% and variation in chronic effects between species is likely to be less than that between acute effects (x is to be defined)

5   That the ratio between acute and chronic effect/no effect responses is likely to be less than 100

6   That concentration (dose)–response relationships can be approximated as linear across critical ranges

7   That the effects of different chemicals tend to impact additively on ecological systems when they are applied in mixtures

1. This is based upon the observation that there is often a considerable amount of structural redundancy in ecosystems, i.e. species can be removed without apparent effect on energy fluxes and material cycles [14, 15]. If this is so then protecting structure should, in general, protect function – this is a kind of ecological precautionary principle.

2. This to some extent counters the notion of structural redundancy by fixing on the fact that certain species may act as keystones within the system (*sensu* Paine [16]); this means that their removal causes profound changes in the structure and functioning of the community of which they are part. Thus, removal of a top predator by eliminating controls on the prey may cause prey populations to increase with knock-on effects for their food. The extent to which this is generally applicable will depend upon the frequency of communities in nature involving keystone species. It seems unlikely, for example, that keystone species will be very prevalent in flowing-water communities where generalist food webs seem to be dominant [17].

Both (1) and (2) depend upon an understanding of the way(s) communities are organized, i.e. the way(s) species contribute to community/ecosystem functioning. Ecological understanding here is not far advanced and this is an area of active current research [18].

3. This is a response to the fact that, in most prospective tests, effects are measured on the survival and reproduction of populations in isolation. Yet indirect effects on trophic and competitive interactions such as those described under (2) may be important. It is also possible that differential toxic effects might affect the balance of predator–prey and competitive relationships, and that bioaccumulation in organisms in one trophic level might influence responses in higher ones. The extent to which direct effects are generally more relevant than indirect ones is likely to depend partially on the relative frequency in nature of precise trophic organizations relative to loosely organized communities with generalist species. On this, the same comments apply for flowing-water communities as in (2) [17].

4. This recognizes, on the one hand, that generally sensitive species do not seem to exist [19] and, on the other, that there have to be limits to the distributions of sensitivities to toxicants within communities. More needs to be known about the form of these distributions if we want to effectively extrapolate from laboratory tests to responses in nature (e.g. [20]). It is also a matter of empirical observation that, between genotypes at least, variations in chronic responses seem to be less than those between acute responses [21,22]. One possible explanation of this is that responses to high doses may be very specific whereas responses to low doses over long periods may be more general. Due to the action of general stressors as selection pressures the latter are likely to converge under the action of natural selection (D.J. Baird, personal communication).

5. This is a response to the fact that risk assessment procedures on new and existing chemicals often involve extrapolation from acute to chronic responses. It is empirical experience that this tends to be less than two orders of magnitude, i.e. a factor of 100 [23], but can this be supported mechanistically? The fact that variance between chronic effects may be less than between acute effects suggests that a generally applicable extrapolation factor may not be feasible (above). Yet, still acute-to-chronic ratios might be expected, in general, to fall within defined limits.

6. This is a response to a common presumption, often implicit rather than explicit, that as ratios of environmental concentrations to effect concentrations reduce there will be a proportional reduction in ecological risk [13]. Yet it is well known that dose (concentration)–response relationships are non-linear. For example, it is often presumed that ecosystems have a capacity to assimilate some levels of toxic chemical but, that beyond a threshold, effects may occur increasingly with increasing concentration to some limit. Hence, an important question is the extent to which a presumption of linearity is appropriate for critical ranges or if it would be more appropriate to use a convenient transformation, the usual one being probit of response versus logarithm of concentration.

7. Additivity of effect in chemical mixtures is a well known general presumption. Empirical observations are increasingly supportive of the assumption [24]. It is also possible to provide a mechanistic explanation of additivity in chemicals that act as general non-polar narcotics [25]. More complex effects are unlikely to be additive. Hence, the extent to which (7) can be applied will vary with the chemicals under consideration.

2.5 CONCLUSIONS

Reliability and relevance are key criteria for the design and interpretation of ecotoxicological programmes. Reliability raises practical questions that need an ecological input for satisfactory resolution. Relevance raises theoretical issues that are at the heart of fundamental ecology. What features do we expect normal ecosystems to possess? Do ecological norms exist, at least in terms of principles that can apply to ecosystems in general? In consequence, these are issues that have not been resolved in ecology so there ought to be an opportunity for fruitful interaction here between the basic and applied science, a point that is emphasized by Kareiva *et al.* [10].

In this context, the 'rules of thumb' that are identified in Table 2.1 are intended to be provocative from both the applied and basic perspectives. On scrutiny they may all turn out to be wrong, but if ecotoxicology is going to develop observational and test systems that are relevant, yet applicable, it will be important to focus on key questions such as these and to consider how they might be addressed.

REFERENCES

1. Underwood, A. J. (1994) Spatial and temporal problems with monitoring, in *The Rivers Handbook* (eds P. Calow and G. E. Petts), Blackwell Scientific Publications, Oxford, Vol. 2, pp. 101–23.
2. Norris, R. H., McElravy, E. P. and Resh, V. H. (1992) The sampling problem, in *The Rivers Handbook* (eds P. Calow and G. E. Petts), Blackwell Scientific Publications, Oxford, Vol. 1, pp. 282–306.
3. Calow, P. (1974) Some observations on the dispersion patterns of two species-populations of littoral, stone-dwelling gastropods (Pulmonata). *Freshwater Biology*, 4, 557–76.
4. Baird, D. J., Soares, A. M. V. M., Girling, A., Barber, I., Bradley, M. C. and Calow, P. (1988) The long-term maintenance of *Daphnia magna* Straus for use in ecotoxicity tests: problems and prospects, in *Proceedings of the First European Conference on Ecotoxicology* (eds H. Lokke, H. Tyle and F. BroRasmussen), Copenhagen, Polytenish Forlag, Odense, pp. 144–8.
5. Calow, P. (ed.) (1993) *Handbook of Ecotoxicology*, Blackwell Scientific Publications, Oxford, Vol. 1.
6. Soares, A. M. V. M. and Calow, P. (eds) (1993) *Progress in Standardization of Aquatic Toxicity Tests*, Lewis Publishers, Boca Raton.
7. Forbes, V. E. and Forbes, T. L. (1994) *Ecotoxicology in Theory and Practice*, Chapman & Hall, London.
8. Calow, P. (1994) Ecotoxicology: what are we trying to protect? *Environmental Toxicology and Chemistry*, 13, 1549.
9. Maltby, L. and Calow, P. (1989) The application of bioassays in the resolution of environmental problems; past, present and future. *Hydrobiologia*, 188/189, 65–76.
10. Kareiva, P., Stark, J. and Wennergen, U. (1996) Using demographic theory, community ecology and spatial models to illuminate ecotoxicology. *This volume*.
11. Calow, P. (in press) Ecosystem health – a critical analysis of concepts, in *Evaluation and Monitoring the Health of Large Scale Ecosystems* (eds D. J. Rapport, C. A. Gaudet and P. Calow), Springer-Verlag, Berlin.
12. Karr, J. R. (1991) Biological integrity: a long-neglected aspect of water resource management. *Ecological Applications*, 1, 66–84.
13. Calow, P. (1993) Hazards and risks in Europe. Challenges for ecotoxicology. *Environmental Toxicology and Chemistry*, 12, 1519–20.
14. Schindler, D. W. (1981) Detecting ecosystem responses to anthropogenic stress. *Canadian Journal of Fisheries and Aquatic Sciences*, 44, 6–25.
15. Gray, J. S. (1989) Effects of environmental stress on species rich assemblages. *Biological Journal of the Linnean Society*, 37, 19–32.
16. Paine, R. T. (1978) Intertidal community structure. Experimental studies on the relationship between a dominant competitor and its principal predator. *Oecologia*, 15, 93–120.
17. Hildrew, A. G. (1992) Food webs and species interaction, in *The Rivers Handbook* (eds P. Calow and G. E. Petts), Blackwell Scientific Publications, Oxford, Vol. 1, pp. 307–30.
18. Lawton, J. H. (1994) What do species do in ecosystems? *Oikos*, 71, 367–74.
19. Cairns, J. (1984) Are single species toxicity tests alone adequate for estimating environmental hazard? *Environmental Monitoring and Assessment*, 4, 259–73.
20. Kooijman, S. A. L. M. (1987) A safety factor for LC_{50} values allowing for differences in sensitivity between species. *Water Research*, 21, 269–76.

21. Baird, D. J., Barber, I. and Calow, P. (1990) Clonal variation in general responses of *Daphnia magna* Straus to toxic stress. I. Chronic life-history effects. *Functional Ecology*, **4**, 399–407.
22. Soares, A. M. V. M., Baird, D. J. and Calow, P. (1992) Interclonal variation in the performance of *Daphnia magna* Straus in chronic bioassays. *Environmental Toxicology and Chemistry*, **11**, 1477–81.
23. Slooff, W., van Oers, J. A. M. and De Zwart, D. (1986) Margins of uncertainty in ecotoxicological hazard assessment. *Environmental Toxicology and Chemistry*, **5**, 841–52.
24. Doi, J.M. (1994) Complex mixtures, in *Handbook of Ecotoxicology* (ed. P. Calow), Blackwell Scientific Publications, Oxford, Vol. 2, pp. 289–310.
25. Donkin, P. (1994) Quantitative structure–activity relationships, in *Handbook of Ecotoxicology* (ed. P. Calow), Blackwell Scientific Publications, Oxford, Vol. 2, 321–47.

3

Using demographic theory,
community ecology and spatial
models to illuminate ecotoxicology

PETER KAREIVA, JOHN STARK AND UNO
WENNERGREN

Ecotoxicology has only weakly addressed fundamental ecological issues because basic ecology itself has only recently begun to provide the necessary tools
for such practical work. Fortunately, this undistinguished record for
ecotoxicology promises to improve as a result of three rapidly advancing
branches of ecological theory. The theoretical innovations that promise to
make ecotoxicology a more predictive science are: (1) applications of stage-
and age-structured demographic models with time-varying rates; (2) using path
analysis to examine cascades of changes in food webs following environmental
perturbations; and (3) using spatially explicit models of population dynamics
to suggest landscape manipulations that could either mitigate or exacerbate
pollutant impacts.

3.1 INTRODUCTION

None of the leading ecology textbooks mention ecotoxicology, even though
environmentalists around the world are well aware of the ecological threats
posed by chemical contaminants. On the other hand, toxicology symposia and
texts make frequent reference to principles of ecology – citing everything from
food webs, to disturbance theory, to ecosystem homeostasis. In practice, however, ecotoxicology is largely toxicology with ecology added as a 'seasoning' as
opposed to a 'main ingredient' – an important applied science dominated by
dose–response curves, vulnerable species assays and reductionist laboratory

ECOtoxicology: Ecological Dimensions. Edited by D. J. Baird, L. Maltby, P. W. Greig-Smith
and P. E. T. Douben. Published in 1996 by Chapman & Hall, London. ISBN hardback
0 412 75470 3 and paperback 0 412 75490 8.

experiments, and bereft of ecological theory [1, 2]. The implication of many critical essays has been that toxicologists are narrow and neglect ecological theory because of intellectual prejudices. While we agree there is a paucity of ecology in ecotoxicology, we suspect the blame lies with the primitiveness of ecological theory (not the failings of toxicologists) and with ecology's inability to provide quantitative tools of the sort needed for environmental risk assessment. Assigning blame, however, does not make for scientific progress. A more interesting question is: can we do better? In spite of a gloomy past record for ecological theory as a predictive tool, we think several recent advances in ecological modelling and theory hold great promise for ecotoxicology. In this chapter we review three major advances in ecological analysis and outline prescriptions for how these advances might usefully be applied in ecotoxicology.

3.2 STAGE-STRUCTURED DEMOGRAPHIC MODELS AS AN IMPROVEMENT ON SIMPLY COUNTING DEAD ANIMALS

To an ecologist, much of the empirical work in ecotoxicology seems to involve little more than tedious tabulations of death rates as a function of chemical concentrations. Obviously, there is a great deal of scientific subtlety to understanding the mechanisms of toxicity, as well as anticipating possible synergy among chemicals – but it is not a subtlety that interests most ecologists. Thus it is easy for ecologists to dismiss a large portion of ecotoxicology as 'irrelevant' or 'uninteresting'. Unfortunately, such a dismissal neglects many of the fascinating demographic issues associated with assessing chemical effects at a population level, issues identical to the applications of demography to the conservation of endangered species, which is a booming area in ecology [3]. It is worth emphasizing that demography becomes important only if the effects of the chemicals under study are significantly less than absolute doom and death. Obviously, if a chemical kills all ages and all life stages with great certainty, demography is a waste of time. However, since most chemicals are diluted or degraded in the environment, there is some point at which their effects become subtle. It is this point of subtle effects that requires demography to answer two questions: (1) how does an organism's rate of population growth or decline change as a function of chemical concentration; and (2) how rapidly can an organism's population recover from brief exposure to toxic compounds that subsequently degrade?

The first question of population change has been addressed by some ecotoxicologists [4–6], who use finite rates of population increase to summarize the effects of chemicals as opposed to simply reporting mortality rates as the complete index of 'toxic effects'. This is not a trivial distinction because it is possible to alter the survivorship rates of some life stages of an organism and yet have negligible impact on population growth [7] – in such cases a risk assessment based on 'mortality in the wrong life stage' (as opposed to population growth) could greatly exaggerate the impact of a chemical.

The issue of population recovery has not, however, been examined using similar demographic models – even though it is ideally suited to such an analysis. The absence of a quantitative focus on population recovery may be explained by the fact that one's first concern sensibly ought to be immediate damage. But, in fact, populations do recover from even the most dramatic of chemical insults (such as oil spills in Alaska) and their rate of recovery is an important facet of environmental risk assessment [8, 9]. The problem is most clearly seen by considering a pesticide that does not kill all individuals in a target population and is not equally lethal to all ages or stages of the target species. Many modern highly specific pesticides fit this description and we have been investigating their impact on pest population recovery using stage-structured demographic models [10].

Toxicologists are well aware that reductions in survival or reproduction due to any chemical can vary dramatically within a species as a function of an individual's age. In order to translate this experimental observation into consequences for population recovery, we have adopted stage-structured population models such as that depicted in Figure 3.1. The basic idea is simple:

$$
\begin{bmatrix}
P_1 & 0 & 0 & 0 & F_1 & F_2 & F_3 & - & - & F_m \\
G_1 & P_2 & 0 & 0 & 0 & 0 & 0 & - & - & 0 \\
0 & G_2 & P_3 & 0 & 0 & 0 & 0 & - & - & 0 \\
0 & 0 & G_3 & P_4 & 0 & 0 & 0 & - & - & 0 \\
0 & 0 & 0 & G_4 & 0 & 0 & 0 & - & - & 0 \\
0 & 0 & 0 & 0 & G_5 & 0 & 0 & - & - & 0 \\
0 & 0 & 0 & 0 & 0 & G_6 & 0 & - & - & 0 \\
- & - & - & - & - & - & - & - & - & 0 \\
- & - & - & - & - & - & - & - & - & 0 \\
0 & 0 & 0 & 0 & 0 & 0 & 0 & 0 & G_n & 0
\end{bmatrix}
$$

Figure 3.1 A life-table matrix as a framework for examining the impacts of chemicals on population growth and recovery. The above matrix is used to iterate populations from one time period to the next with the timescale chosen to match the time intervals on which data are collected. The parameters in the matrix change with chemical exposure and by multiplying the matrix several times one can summarize all of these effects into a portrait of population growth or population recovery. σ_i, Survivorship during stage i (from life-table data); γ_i, developmental rate in stage i (from life-table data); F_i, fecundity during age-class i as adult (from life-table data). $P_i = \sigma_i(1 - \gamma_i)$, probability of surviving and remaining in the same stage. $G_i = \sigma_i\gamma_i$, probability of surviving and growing to the next stage-class ($\gamma_i = 1$ in the adult stage since the time step is equal to the age-class interval).

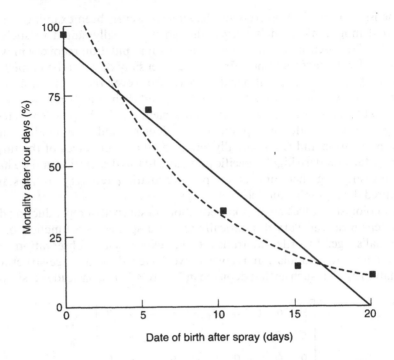

Figure 3.2   The change in toxicity of Margosan-O as measured by survival of newborns never directly treated with insecticide, but placed on plants that had been sprayed 5, 10, 15 and 20 days previously with Margosan-O at a rate of 100 ppm azadirachtin. Experimental details are given in [11]. —, Linear curve; - - -, half-life curve; ■, experimental data point.

one divides the target population into stages (such as newborn, juveniles and adults) or ages (e.g. in terms of days for aphids), and empirically determines how the fecundity, survival and growth rates vary as a function of insecticide application. Elsewhere we report in detail the results of such experiments, in which Margosan-O (a naturally derived insecticide from the Neem tree in which the major active ingredient is azadirachtin) was applied to cohorts of aphids of different ages [11]. Of course, the daily rates summarized by a matrix such as that given in Figure 3.1 will vary through time as a function of residual activity of a pesticide – but the temporal decay in this activity can also be directly estimated (see Figure 3.2).

By piecing together age- or stage-specific demographic impacts and decay in pesticide activity, it is straightforward to ask how the initial age structure of a target population and the details of pesticide persistence interact to determine population recovery. To summarize what would otherwise be a bewildering array of population trajectories we have adopted two indices of population recovery: (1) long-run intrinsic rate of increase assuming a birth and mortality matrix which is constant through time; (2) the number of days a sprayed population's growth will lag behind an unsprayed population's. The long-run rate of population growth allows us to summarize a treatment effect (which includes changes in mortality, birth rate and growth rate) with one ecological relevant number. The delay in population growth index allows us to include time-varying changes due to insecticide activity gradually decaying and has the practical value of quantifying how many days of crop growth would be gained from an insecticide treatment.

Using these measures, and a stage-structured demographic model, several subtleties of the effect of Margosan-O on aphid populations become clear. Firstly, the age structure of the aphid population during a spray application is enormously important, with rates of population growth following insecticide treatment varying by as much as threefold depending on the prevalence of newborn versus adult aphids (Figure 3.3). Secondly, even with modest concentrations of Margosan-O, consideration of the insecticide's persistence in the environment easily doubles the delay in aphid recovery compared to a scenario that neglects the insecticide's persistence (Figure 3.4). Elsewhere we

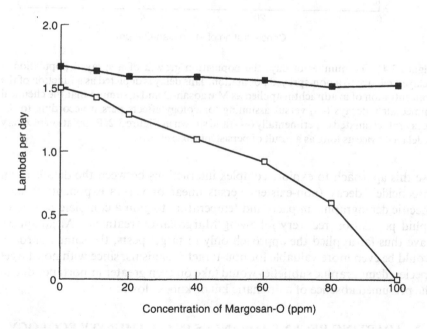

Figure 3.3 The asymptotic rate of population change (lambda) for a population of aphids exposed to various concentrations of Margosan-O, assuming either juveniles (□) or adults (■) are targeted. Lambda is the multiplication factor for the population on a daily basis, such that a lambda greater than one implies exponential growth and a lambda less than one implies exponential decline. Clearly, exposing adults makes little difference, whereas population growth rates can be reduced to zero when juveniles are exposed.

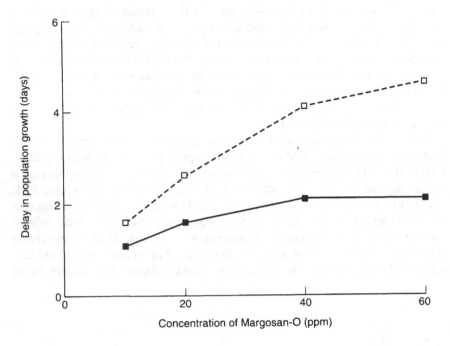

Figure 3.4 The number of days the population growth of a sprayed population is delayed relative to a non-sprayed population. This delay is depicted as a function of the concentration of azadirachtin applied as Margosan-O and assuming that the chemical immediately decays (—) versus assuming toxicology effects persist according to the decay rate estimated experimentally (- - -) and shown in Figure 3.2. Population recovery is delayed twice as long as a result of persistent toxic effects.

use this approach to explore complex interactions between the details of an insecticide's decay (non-existent versus linear or versus exponential), age-specific demographic impacts and temperature to gain a complete picture of aphid population recovery following Margosan-O treatment. Although we have thus far applied the approach only to target pests, the same approach would be even more valuable for non-target organisms, since with non-target species, demographic subtleties would take on even greater importance due to the presumed absence of a dramatic kill or knock-down.

3.3 ADAPTING RECENT ADVANCES IN COMMUNITY ECOLOGY FOR PREDICTING BEYOND SINGLE-SPECIES RESPONSES IN ECOTOXICOLOGY

Several ecologists have suggested that the 'single-species' approach to toxic-ology is inadequate (e.g. [12]). The solution most often touted is the use of multispecies 'mesocosms', in which total community responses might be directly observed [13]. While multispecies mesocosms are practical for microbial systems and many aquatic systems, their use in terrestrial communities is virtually non-existent. The challenge then is to ask whether it would be possible to explore the cascade of effects due to a chemical using field studies of species-rich terrestrial communities. Two recent pioneering studies in ecology offer suggestions for how this could be accomplished.

Cohen *et al.* [14] used a unique data set on food web connections in rice ecosystems to ask how pest population dynamics, as mediated by food web connections, were altered by repeated applications of the pyrethroid, deltamethrin. The rice ecosystem is especially fascinating because it includes a web of natural enemies that is remarkable in its diversity and in the richness of its connections. In particular, over 645 species depend on rice and its herbivores for their living, with over 9000 trophic links (consumer–resource links) firmly documented [14]. Each of the major pest insects in rice seem to be strongly affected by anywhere from three to five natural enemies, with the suite of controlling enemies varying between the rice pests. Using repeated censuses of seven major rice pests and their natural enemies (taken throughout the growing season along with measures of rice biomass), Cohen *et al.* asked what proportion of the temporal variation in pest densities could be explained by rice biomass, past pest densities, natural enemy densities and the presence or absence of insecticide itself. One finding was that insecticides did not consistently yield reductions in pest density compared to controls because of presumed impacts on enemies. Especially intriguing was the impact of insecticides on the 'predictability' of pest populations as measured by the percentage variance explained using the best possible multiple regression model. Spraying the rice with pyrethroids disorganized pest dynamics such that if a field had been sprayed it became increasingly difficult to forecast pest populations using a wide variety of independent variables. In particular, the average r^2 for regression models describing pest population fluctuations was only 0.51 in the presence of pesticides compared to 0.68 in unsprayed fields –this meant that a significant portion of the variation in pest density could be predicted by a simple regression model for four of seven pest species as long as no pesticides were sprayed, but such 'predictability' was possible for only one of the seven pest species once the rice ecosystem was subjected to insecticides. The idea that a chemical could disorganize population dynamics is something that could never be detected from laboratory toxicological assays, but it could be terribly relevant to a farmer who seeks stability from year to year.

An alternative to Cohen *et al.*'s approach for multispecies systems is provided by Wootton [15] who introduced the statistical technique of 'path analysis' to studies of ecosystem dynamics. Here, instead of a regression model, one draws a web describing pathways of influence following an application of any environmental perturbation (species removal, species addition, chemical application, etc.). This diagram of connections is then examined using partial correlation matrices to come up with all possible direct and indirect effects –

each of which can be given a relative importance value and a significance value [16]. The beauty of this approach is that it allows one to separate the direct effects of a toxicant (by killing a pest) from indirect effects (by killing individuals that feed on the pest). For example, one can write the heuristic model:

$$SD = a_{TS} T + a_{PS} PD$$
$$PD = a_{TP} T$$

where SD denotes the decline in spider abundance, PD denotes the decline in abundance of spider prey resources, T is the concentration of an insecticide application, a_{TS} and a_{TP} depict the correlation between insecticide concentration and direct killing of spiders or their prey, and a_{PS} denotes the correlation between declines in prey abundance and declines in spider abundance. By formally estimating the 'a_{ij}' coefficients above, one can quantify the relative importance of direct and indirect effects of some toxicant [17]. Of course, the above illustration is simplistic and the real value of path analysis emerges with more complex webs of interaction. In general, Cohen *et al.*'s and Wootton's innovations for multispecies studies represent a different direction to much previous multispecies testing in ecotoxicology. Specifically, rather than simply asking how many different species might suffer enhanced mortality, they ask how population dynamics or webs of interactions might be altered. If one is concerned with ecosystem function, this is the real question – not simply counting dead animals of many different species.

3.4 USING SPATIALLY EXPLICIT MODELS TO SUGGEST HOW THE SCALE AND SPATIAL PATTERNING OF A CHEMICAL STRESS INFLUENCES ITS IMPACT

Ecologists pride themselves on 'studying nature', which to many means 'the great messy, unconfined outdoors' – free of the shackles of a laboratory environment. In contrast, much of ecotoxicology relies on controlled laboratory experiments – largely because dosages and concentrations have to be precisely controlled if a predictive hazard assessment is a goal. Yet, the effects of any perturbation – whether it is chemical, physical or biological – depend on the spatial scale of the event, the composition of the surrounding landscape and the dispersal rates of the organisms affected by the perturbation [18]. In some senses ecotoxicology is far ahead of ecology in dealing with this effect because some enlightened ecotoxicologists have experimentally assessed the consequences of perturbations (insecticide spraying) carried out at different scales [19], whereas ecologists have largely just talked about the importance of such experiments [20]. However, fast-breaking developments in ecological theory provide precise guidance for how spatial scale can be crucial to any examination of toxic effects at the ecosystem level. In particular, a few robust abstract generalities have emerged from spatially explicit models that are worth tailoring to questions about ecotoxicology. For instance, models of interacting species (especially consumer–victim linkages) suggest that there is some intermediate level of dispersal for species that will dampen the propagation in time and space of local disturbances [21]; since land-use management determines animal dispersal (e.g. whether a landscape is all farmland or is a checkerboard of woodlands and crops), there is thus the opportunity to manipulate landscapes in a way that would minimize the unwanted effects of insecticide applications. A related robust finding from spatially explicit models is that the size of a landscape in which interacting organisms move about and are exposed to perturbations governs population stability and fluctuations – with small landscapes or arenas generally being the most unstable [22]. Again, this abstract idea may bridge to the practical world of ecotoxicology by suggesting different effects for chemicals depending on the size of continuous patches of habitat into which a given chemical is introduced. Even if ecotoxicologists find these spatially explicit models too fanciful, ecological theoreticians ought to be attracted to ecotoxicology as a tool for testing their models. Chemical applications represent well-defined and quantifiable perturbations whose subsidiary effects on population dynamics could be explicitly examined from the perspective of spatial models (in contrast, most such theory expresses 'perturbation' or 'disturbance' in the most vague of terms, with no concrete event in mind).

3.5 CAN ECOLOGY BETTER SERVE ENVIRONMENTAL TOXICOLOGY?

There never has been, and probably never will be, an ecological society meeting devoted to 'issues in toxicology', even though much of the natural world is altered by humanmade chemicals. In contrast, toxicologists readily turn to ecology for ideas. Unfortunately, when they do so, they find very little that will help them predict long-term and large-scale consequences of chemical perturbations. The reason is simple – basic ecology has been slow to develop the tools needed by ecotoxicologists. The three areas we discuss (demographic models, multispecies studies and spatially explicit dynamics) provide excellent examples of challenges that basic ecology has only recently addressed. For example, while matrix models of population growth were introduced half a century ago, their application to field population dynamics involves only a handful of papers and those studies that do use demographic matrices typically assume matrices that are invariant through time [23]. The study of multispecies systems in basic ecology lags even farther behind. For instance, a review of all papers published in the journal *Ecology* (the leading journal of the Ecological Society of America) from 1980 to 1990 revealed that only one in five papers even mentioned more than four species [24]. Yet, even in the simplest agricultural community, it would be impossible to imagine a chemical insecticide that affected the abundances of only four species. Since basic research in ecology has ineffectively dealt with multispecies systems, it is no surprise that ecotoxicology is mired in the single-species approach. Finally, the limitations of small-scale experiments haunts basic ecology at least as much as it does ecotoxicology. Indeed, over half of the studies of species interactions and population dynamics that were published in *Ecology* between 1980 and 1986 spanned distances no larger than one metre [25]! Only within the last few years have ecologists begun to ask whether extrapolations from experiments conducted in tiny field plots should be trusted if predictions are sought at larger scales.

3.6 CONCLUSION

It appears to us that the absence of 'ecology' in 'ecotoxicology' is partly due to the contrived domain of ecology, in which the basic science that calls itself ecology has only recently begun to address the dynamics of structured populations in large-scale landscapes and involving interactions among dozens of species. But now that ecologists are finally putting the 'eco' into ecology, the time is ripe for putting ecology into ecotoxicology. Not only will ecotoxicology benefit from this infusion, but the basic science of ecology will be forced to sharpen its tools and focus its theories as it deals with applied questions that demand quantitative answers, and not just the reporting of this or that 'significant effect' in some analysis of variance table.

ACKNOWLEDGEMENTS

This work was supported by an EPA grant for Exploratory Research to P. Kareiva, by a USDA grant to John Stark and by a grant from the Swedish Wenner-Grer Center Foundation. We especially thank Professor R. T. Paine's garden for inspiration regarding aphid pests, Professor Paine himself for indulging the interests of wayward underlings, and the director of WSU Experiment Station for tolerating John Stark's whimsies.

REFERENCES

1. Cairns, J. Jr (1989) Editorial: Will the real ecotoxicologists please stand up? *Environmental Toxicology Chemistry*, **8**, 843–4.
2. Cairns, J. Jr and Pratt, J. (1993) Trends in ecotoxicology. *The Science of the Total Environment*, Supplement 1993, 7–22.
3. Schemske, D. W., Husband, B. C., Ruckelshaus, M. H., Goodwillie, C., Parker, I. and Bishop, J. G. (1994) Evaluating approaches to the conservation of rare and endangered plants. *Ecology*, **75**, 584–606.
4. Ahmadi, A. (1983) Demographic toxicology as a method for studying the two-spotted spider mite system. *Journal of Economic Entomology*, **76**, 239–42.
5. Allan, J. and Daniels, R. (1982) Life table evaluation of chronic exposure of *Eurytemora affinis* to kepone. *Marine Biology*, **66**, 179–84.
6. Bechmann, K. R. (1994) Use of life tables and LC_{50} tests to evaluate chronic and

acute toxicology effects of copper on a marine copepod. *Environmental Toxicology & Chemistry*, **13**, 1509–17.

7. Crouse, D. T., Crowder, L. B. and Caswell, H. (1987) A stage-based population model for loggerhead sea turtles and implications for conservation. *Ecology*, **68**, 1412–23.
8. Cairns, J. Jr (1980) *The Recovery Process in Damaged Ecosystems*, Ann Arbor Science Publishers, Ann Arbor, MI.
9. Leppakoski, E. and Lindstrom, L. (1978) Recovery of benthic macrofauna from chronic pollution in the sea area off a refinery plant, southwest Finland. *Journal Fisheries Research Board Canada*, **35**, 766–75.
10. Caswell, H. (1989) *Matrix Population Models*, Sinauer Press, Sunderland, MA, USA.
11. Stark, J. and Wennergren, U. (in press) Can population effects of pesticides be predicted from demographic toxicological studies? *Journal of Economic Entomology*.
12. Cairns, J. Jr (1983) Are single-species tests alone adequate for estimating hazard? *Hydrobiologia*, **100**, 45–57.
13. Taub, F. (1989) Standardized aquatic microcosm – development and testing, in *Aquatic Ecotoxicology* (eds A. Boudou and R. Ribeyre), CRC Press Inc., Boca Raton, FL, Vol II, pp. 47–92.
14. Cohen, J. E., Schoenly, K., Heong, K. L., Justo, H., Arida, G., Barrion, A. T. and Litsinger, J. A. (1995) A food-web approach to evaluating the effect of spraying on insect pest population dynamics in a Philippine irrigated rice ecosystem. *Journal of Applied Ecology*, **31**, 747–63.
15. Wootton, J. (1994) Predicting direct and indirect effects: an integrated approach using experiments and path analysis. *Ecology*, **75**, 151–65.
16. Mitchell, R. (1992) Testing evolutionary and ecological hypotheses using path analysis and structural equation modeling. *Functional Ecology*, **6**, 123–9.
17. DeAngelis, D. L. (1996) Indirect effects: concepts and approaches from ecological theory. *This volume*.
18. Jepson, P. and Sherratt, T. (1996) The dimensions of space and time in the assessment of ecotoxicological risks. *This volume*.
19. Jepson, P. C. and Thacker, J. R. M. (1990) Analysis of the spatial component of pesticide side-effects on non-target invertebrate populations and its relevance to hazard analysis. *Functional Ecology*, **4**, 349–55.
20. Levin, S. (1992) The problems of pattern and scale in ecology. *Ecology*, **73**, 1943–67.
21. Kareiva, P. (1991) Population dynamics in spatially complex environments: theory and data. *Philosophical Transactions of the Royal Society of London, B*, **330**, 175–90.
22. May, R. (1994) The effects of spatial scale on ecological questions and answers, in *Large Scale Ecology and Conservation Biology* (eds P. Edwards, R. May and N. Webb), 35th Symposium of the British Ecological Society, Blackwell Scientific Publications, Oxford, UK, pp. 1–19.
23. Doak, D., Kareiva, P. and Klepetka, B. (1993) Modelling population viability for the desert tortoise in the Western Mojave desert. *Ecological Applications*, **4**, 446–60.
24. Kareiva, P. (1994) Higher order interaction as a foil to reductionist ecology. *Ecology*, **75**, 1527–8.
25. Kareiva, P. and Anderson, M. (1988) Spatial aspects of species interactions: the wedding of models and experiments, in *Community Ecology* (ed. A. Hastings), *Lecture Notes in Biomathematics 77*, Springer-Verlag, Berlin, Germany, pp. 35–50.

4   *Indirect effects: concepts and approaches from ecological theory*

DONALD L. DeANGELIS

Indirect effects are changes in the abundance of a population resulting not directly from the action of a causal agent (such as a toxicant) but indirectly through the effects of the causal agent on other species. Examples are presented of simple indirect effects that have been studied by theoretical ecologists. The possible use of both empirical studies and ecosystem models to predict indirect effects of one species on another is discussed. Both approaches have been used to estimate the relative magnitudes of direct and indirect effects in food webs. However, the uncertainty of interspecific interactions may multiply when indirect effects are computed, leading to low confidence in predictions of effects. Environmental and demographic stochasticity, as well as population density-dependent self-regulation, may diminish the influence of indirect effects.

4.1   INTRODUCTION

Perhaps the most well known effects of toxicants on wildlife are those resulting from movement of toxicants through the food web to high level consumers, where they have negative effects on populations of these consumers. For example, DDT and other organochlorine pesticides, used to kill insects, become concentrated in the tissues of birds that feed on the insects and lead to such physiological effects as eggs with abnormally thin shells that crack during incubation [1]. These indirect effects can be predicted if there is information on the trophic pathways of food webs and the physiological mechanisms behind toxicant action on organisms in the food chain.

ECOtoxicology: Ecological Dimensions. Edited by D. J. Baird, L. Maltby, P. W. Greig-Smith and P. E. T. Douben. Published in 1996 by Chapman & Hall, London. ISBN hardback 0 412 75470 3 and paperback 0 412 75490 8.

The effects described above are commonly referred to as secondary poisoning and will be considered 'direct effects' in the rest of this chapter because the reductions in populations of the higher consumers are caused directly by the toxicant, even if the transmission of the toxicant takes place through one or more intermediate species populations.

The focus of this chapter is on indirect effects of toxicants on wildlife. Indirect effects are less widely appreciated, and less easily predicted, and are transmitted to a species population through a series of changes in population levels of one or more intermediate species. For example, O'Connor and Shrubb [2] point out that populations of linnet (*Carduelis cannabina*) have declined in cereal producing regions of Great Britain because of the disappearance of many of its main food plants due to the increase in herbicide use. The same authors note that the reed bunting (*Emberiza schoeniclus*) has probably declined as a result of a loss of nest site habitat through herbicide destruction of weeds in clover fields. Potts [3] found that grey partridge chick survival declined in agricultural areas with increasing levels of pesticide application. Aquatic examples are also widespread. Pesticides applied to farm fields that drain into waterbodies may negatively affect fish populations by decreasing aquatic insect and other invertebrate prey.

Another way in which these 'density-mediated indirect effects' have come to the attention of researchers is through the attempt to use population monitoring as an indication of the presence of chemical contaminants in the environment. According to the rationale for this approach, if a toxicant increases the mortality rate of a population, or decreases reproductive success, this should be reflected in decreases in the population. However, a review of publications spanning a 40 plus year period on population studies of small mammals indicated that such studies rarely yielded definitive results [4]. Indirect population effects appear to explain these apparent inconsistencies.

As an example, Barrett and Darnell [5] sprayed dimethoate, an organophosphate insecticide, on a clover field to assess effects on small mammal populations. Intensive measurements before and after spraying revealed that the insecticide had no direct effects on small mammals but indicated that populations of house mouse (*Mus musculus*) declined, populations of prairie vole (*Microtus ochrogaster*) increased and populations of prairie deer mouse (*Peromyscus maniculatus*) remained stable. These observations were explained as results of indirect effects. Reduction of insect abundance caused a reduction through emigration of *Mus musculus*. *Microtus ochrogaster*, primarily a herbivore, moved into areas no longer defended by the aggressive *Mus musculus*. This example illustrates how toxic chemicals may sometimes appear to benefit certain species.

Another example is the observed effect of sludge treatment (containing nutrients but also heavy metals) on small mammals in an old field [6]. Survivorship of meadow voles (*Microtus pennsylvanicus*) was seen to decline following the treatment. Moreover, application of fertilizer in addition to sewage sludge resulted in an even greater decline in meadow vole populations. Here again, the occurrence of indirect effects can be argued to explain the observed effects. The nutrient, both in the sludge and fertilizer applications, caused a reduction in the diversity and availability of edible plant species which led to a decrease in the population of voles.

It appears from the above examples that indirect effects mediated through changes in population levels are common enough that they must be taken seriously by ecotoxicologists. The study of these types of indirect effects is outside of the usual domain of most ecotoxicologists but is a subject of increasing interest to ecologists. The remainder of this chapter discusses some of the progress made in understanding indirect effects in ecological systems.

4.2   SIMPLE TYPES OF INDIRECT EFFECTS

The study of ecology includes the myriad interrelationships that can occur among different organisms. Despite the principle that every member of an ecological community ultimately affects every other member, systematic studies of interactions have generally focused on two-species interactions. These include predator–prey interactions where one species is affected positively and one negatively (+, –), competitive interactions in which both species suffer (–, –) and mutualistic interactions in which both species benefit (+, +).

The traditional emphasis on pairwise interactions is because they are easiest to study in the field and easiest to analyse using mathematics. As ecologists attempted to explain community patterns, however, it became evident that interactions involving more than two species are often crucial. One of the most notable examples is that of 'keystone predation'. Paine [7, 8] noted in the rocky intertidal zone of northwest USA that competition for space between sessile organisms could lead to eventual exclusion by one species (*Mytilus californicus*) of all others, but the presence of a predator, *Pisaster ochraceus*, preferring the dominant, could prevent its monopolization of space. Such keystone predators have been found in other systems as well (e.g. [9]). Keystone predation is a type of indirect effect because the predator, P, while not directly affecting competitor species, N_2, can facilitate its survival by predation on species N_1, which is a competitor of species N_2. It makes sense to think of this as a basic three-species interaction, depicted abstractly in Figure 4.1a. Any change in density of the predator will have an indirect effect on species N_2.

We can introduce some general definitions of direct and indirect effects at this stage (based on [10]).

1. **Direct effect.** Direct effects are those in which a change in a 'receiver species' (B), caused by a change in an 'initiator species' (A), does not involve other species.
2. **Indirect effect.** An indirect effect occurs when species A (the initiator species) produces a change in species B (the receiver species) by affecting one or more others (transmitter species).

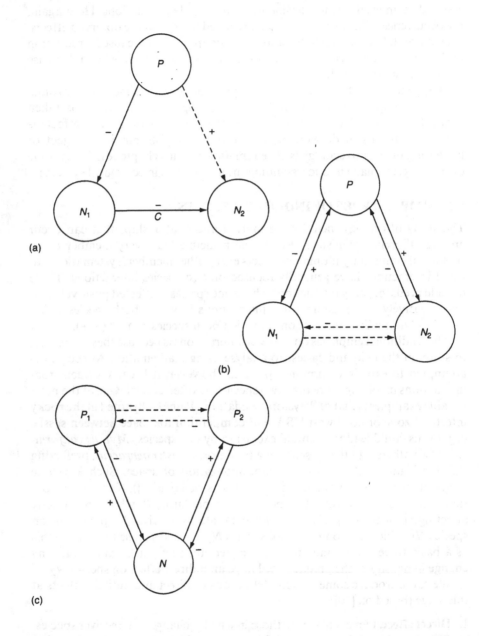

(a)

(b)

(c)

Figure 4.1 Schematic diagrams of important indirect effects. (a) Keystone predation; (b) apparent competition; (c) exploitation competition. The solid lines represent direct effects, while the dashed lines represent the indirect effects. The symbol P represents a predator population, N represents a prey population and C represents a competitive effect.

When we apply the definition of indirect effect to ecotoxicology, the 'initiator' is a toxicant. Then we are concerned with its effects on some species A, transmitted to B through direct effects on one or a set of species, C.

Another type of three-species interaction leading to an indirect effect was noted by Holt [11, 12]: 'apparent competition' (Figure 4.1b). This is another case of two prey and one predator, but in this case there are direct effects of the predator on the two prey but no direct competitive effects between the prey. Here the indirect effect is the effect of one prey on another. An increase in the population abundance of prey N_1 stimulates an increase in the predator, which causes a decrease in the abundance of prey N_2. Therefore, there is an 'apparent competition' between prey species N_1 and N_2, even though they do not directly compete.

A third type of indirect effect is 'exploitation competition'. In this case there are two consumers utilizing a single resource. This can be two autotroph species using the same limiting nutrient, two grazer species feeding on the same autotroph and so forth. The two consumers are not pictured as having direct effects on each other (Figure 4.1c) because they are assumed not to have direct interactions, such as competition for space or agonistic encounters over resources. However, each has a negative effect on the resource. Thus, an increase in the abundance of consumer P_1 will have an indirect negative effect on the abundance of consumer P_2 by decreasing the level of the mutual resource, N.

If the trophic chain is extended beyond two levels to three or more levels, a type of indirect effect termed a 'trophic cascade' can occur (Figure 4.2a). Suppose that each consumer in this chain is able to exert a negative effect on the population of the species below it. If there is an increase in the abundance of the top-level species, this will have an indirect positive effect on the bottom species in this three-trophic level chain. This trophic cascade effect can be extended to longer chains as well (e.g. Figure 4.2b). The trophic cascade concept has been employed in lake ecosystems as a way of providing biological control of undesirable levels of algal abundance [13]. By stocking piscivorous fish in a lake, managers can attempt to reduce planktivorous fish, which may have the effect of increasing zooplankton, which can reduce algal populations.

Important indirect effects that depend on four-species interactions also occur. In particular, the type of indirect effect termed 'indirect mutualism' has been noted in various systems. In the rocky intertidal of the Gulf of California, the gastropod *Acanthina angelica* preys on the barnacle *Chthamalus anisopoma*. The limpets *Collisella strongiana* graze on algae species (*Ralfsia* sp.). The barnacles and algae compete for space (Figure 4.3). An increase in either of the consumers, either the predatory gastropod or the grazing limpet, will have a positive effect on the other consumer. To see this, note that an increase in *Acanthina* will decrease its *Chthamalus* prey, allowing expansion of the space occupied by *Ralfsia*, which in turn permits an increase in *Collisella*.

Another example of indirect mutualism was observed by Brown *et al.* [14] in studies on competition between rodents and other graminivores at Sonoran

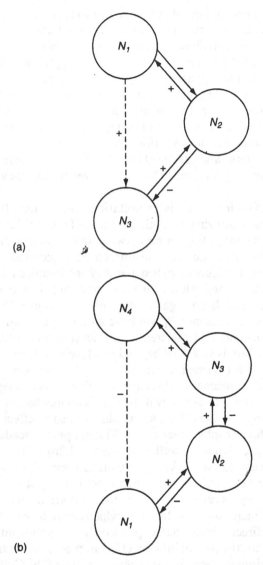

Figure 4.2 Schematic diagrams of indirect effects. (a) Three-species trophic cascade; (b) four-species trophic cascade. The subscript represents trophic position in the food chain.

desert sites. Short-term experiments indicated competition between rodents and ants because they both exploited seeds (Figure 4.4a). However, in experiments that were conducted over longer periods of time, ants and rodents tended to have positive effects on each other. This is explained as follows. The ants specialized on small seeds and the rodents on large seeds. Hence, high ant

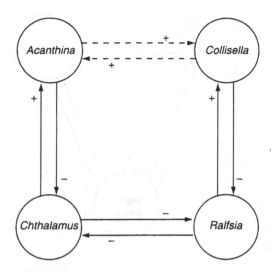

Figure 4.3 Schematic diagram of indirect mutualism. There are direct competitive effects between the barnacles (*Chthamalus*) and algae (*Ralfsia*), each of which has an important consumer – the gastropod *Acanthina* for the barnacles and the limpet *Collisella* for the algae. An indirect mutualism exists between the limpets and gastropods because an increase in, say, limpets, reduces algae, which favours barnacles and thus gastropods, and vice versa.

populations reduced the numbers of small-seeded plants, allowing higher densities of the competing large-seeded plants that supported rodent populations. Similarly, in experimental manipulations in which the rodents had high densities, this favoured the increase of small-seeded plants, which benefited the ant species.

The above examples are all cases of what have been defined as 'density-mediated' indirect effects [10]. These are classified as cases in which the transmission of the effect through intermediate species is due to changes in the population density of the intermediate species. Another general type is called the 'trait-mediated' indirect effect [10]. In this type the transmission of the effect through intermediate species is due to a change in behavioural, physiological or morphological attributes of the intermediate species. As an example, consider the trophic cascade pictured in Figure 4.2. The top consumer is pictured as having a negative influence on the abundance of the intermediate consumer. However, it is also feasible that the top consumer affects the behaviour rather than the abundance of the intermediate species. Suppose, for example, that the intermediate consumers alter their behaviours to avoid their predators. This may alter their population size, but even if it does not, it may alter their influence on the lowest level species, N_1. Trait-mediated indirect effects have not received much attention [10], but a number of documented

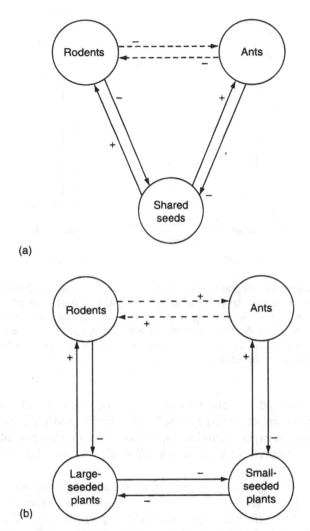

Figure 4.4 Schematic diagram of indirect interactions between desert rodents and ants exploiting seed resources. (a) Over the short term the interaction is exploitative competition; (b) over the long term an indirect mutualism is apparent between the rodents and ants, which compete for seeds of competing plant species.

studies by Lima and Dill [15] show that changes in foraging rate and prey selection in a consumer may be caused by the presence of a predator on that consumer.

The examples above do not exhaust the basic types of indirect effects. In fact, Menge [16], in a survey of rocky intertidal systems, identified nine basic types of indirect effects, with 83 subtypes, including sequences involving up to six species.

These indirect effect types described above are simple enough for one to understand and make predictions. Ecological systems in nature contain far more species, however, so the results from these simple systems are only useful if the natural system can be approximated as being composed of a number of tightly coupled simple subsystems with weak links between these subsystems. This may often be a reasonable assumption. Nevertheless, one needs to develop terminology and techniques for the study of more complex interactions.

In large systems of interacting species, not only can there be longer chains of indirect effects from one species to another, but there may be many different pathways that the indirect effects can follow. Figure 4.5a illustrates this situation. The number of possible pathways of influence from species A to species B is nine in this case (assuming each pathway can go through each node only once). We can term the effect of one particular pathway the 'individual

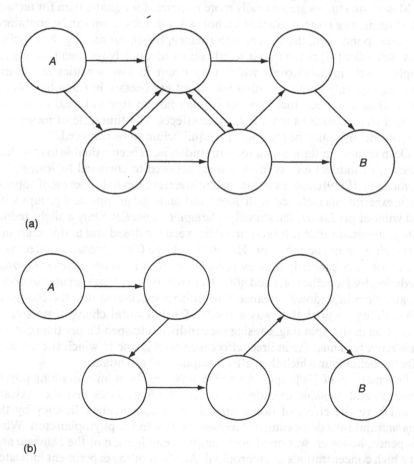

(a)

(b)

Figure 4.5 Schematic diagram depicting (a) the net effect of indirect effects from *A* on *B*; (b) an individual indirect effect.

effect' of that pathway (Figure 4.5b). The sum of the indirect effects of all pathways is termed the 'net effect'. Later, we will discuss mathematical techniques for calculating the net effect in a complex system.

4.3 EMPIRICAL STUDIES TO DETERMINE INDIRECT EFFECTS

Ecotoxicologists have sought to go beyond the single-species toxicity tests that have long been used to evaluate the potential hazard of chemicals in the environment. To determine indirect effects that can occur as a result of toxicant exposure, experiments that simulate natural systems are needed. Because field studies are expensive, outdoor mesocosms that attempt to mimic the natural environment have frequently been used. Besides being less expensive, these systems are also easier to manipulate and replicate than field studies.

Mesocosm studies are generally more practical for aquatic than for terrestrial systems, as a large amount of a whole aquatic ecosystem can be contained in a small pond. Still, there is considerable scepticism concerning their usefulness. Schindler [17] noted that it is difficult to adequately represent the higher trophic levels in mesocosms which may occur in low densities in natural systems, e.g. only about ten adult lake trout per hectare in Canadian Shield lakes. Thus, it is clear that mesocosms may not be able to reveal important indirect effects, such as trophic cascade effects. The timescale of mesocosm experiments may also be too short for equilibrium to be achieved.

Despite their limitations, mesocosm studies have been valuable in revealing a variety of indirect effects in food webs exposed to chemical toxicants [18]. Clements *et al.* [19] exposed stream macroinvertebrates to low levels of copper. Their experiments included both dosed and undosed groups, and groups with and without predators, the stonefly (*Paragnetina media*). Only a slight reduction of stream invertebrates was noted between the dosed and undosed groups in the absence of the predator. However, a large Cu × predation effect was noted for two caddisfly species (*Chimarra* sp. and *Hydropsyche morosa*). Predation by stoneflies on caddisflies was two to three times greater in dosed streams than in undosed streams. The authors speculated that the change in vulnerability to predation was a result of behavioural changes, possibly a disruption in silk-spinning, causing the caddisflies to spend more time outside their retreats. Thus, the indirect effect seen here is one in which the toxicant affects a behaviour which then affects population abundance.

Borgmann *et al.* [20] exposed a stable 3400 dm^{-3} system containing phytoplankton and *Daphnia* to cadmium to test the hypothesis that the toxicant would have the effect of decreasing biomass conversion efficiency by the *Daphnia* and thus decreasing its biomass relative to the phytoplankton. What happened, however, was an almost complete elimination of the *Daphnia* and very high concentrations of chlorophyll. Analysis of the experiment indicated that the effect of cadmium caused *Daphnia* to decrease and phytoplankton to increase to the point at which the algae began to inhibit *Daphnia* growth (an effect that had been noted previously). This led to a positive feedback cycle that eliminated *Daphnia*. The authors note that in a natural system this would have led to establishment of other zooplankton species.

Taub *et al.* [21] exposed a *Daphnia*–phytoplankton population to malathion for 12 hours. The maximum density reached by the *Daphnia* was higher in the treatment than in the control. This was interpreted to be a result of indirect effects. In the dosed system, malathion slowed the growth of the *Daphnia* which allowed the phytoplankton biomass to reach a higher level than it would have otherwise. The *Daphnia* response to the increased phytoplankton exceeded the inhibitory effect of malathion, so the *Daphnia* reached a higher peak in the treated case. Of course, this is an effect of the system being away from equilibrium which might not hold over a period extended enough for the system to approach a steady state.

Finally, Kautsky and colleagues at the University of Stockholm have used experimental rock pool systems to study exposure to cadmium. They have found a number of indirect effects resulting from the reduction of *Daphnia*, including increased phytoplankton and primary production, increased pH values and decreased nutrient concentrations.

4.4 ESTIMATING MAGNITUDES OF DIRECT AND INDIRECT EFFECTS

There is much interest in estimating the magnitudes of indirect versus direct effects in food webs. This interest is partly purely intellectual. Ecosystems are regarded more and more from the point of view of 'complex systems' whose study must involve understanding complex chains and loops of cause and effect.

There are two basic methods that can be used to compare the magnitudes of direct and indirect effects [10]. The first method (the 'theoretical' method) is to study the ecological system in detail through observation to obtain a picture of the general structure of the system (see Figure 4.6 for a hypothetical example) and through short timescale experiments and construct a mathematical model. The short timescale population density perturbation experiments (i.e. 'pulse' experiments, see [22]) will give the direct effects of one species on another denoted by the f_{ij} (*,*)s in Figure 4.6. Mathematical techniques are then used to calculate the indirect effects that can only manifest themselves over a long timescale (e.g. effects of chains such as $X_i \rightarrow X_j \rightarrow ... \rightarrow X_k$), sum over all of these indirect effects and compare the magnitudes of the direct and indirect effects [23].

The second method (the 'experimental' method) is to manipulate densities of populations in natural or mesocosm systems over longer periods of time (i.e. 'press' experiments, see [22]) and measure the changes that eventually take place in other populations in the food web. The direct and indirect effects are sorted out on the basis of the knowledge and observations of the experimenter.

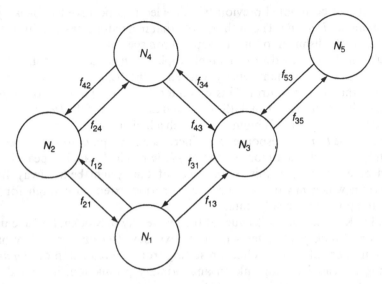

Figure 4.6 Schematic diagram depicting direct interactions among a set of species (N_1–N_5), showing the interaction strengths, f_{ij}.

Neither of these approaches is ideal. The experimental approach has the disadvantage that it may take a very long period of time for all indirect effects to manifest themselves. Yodzis [24] has shown that it may be necessary to keep track of effects up to dozens of links in length to include the majority of the magnitude of indirect effects in a relatively complex food web. He analysed a 30-component marine ecosystem food web. Among his findings were that there are 2 721 747 distinct pathways of seven links or fewer from seals to hake; there are 28 722 675 of eight links or fewer. Yodzis also found that one could not normally obtain a good approximation of the net effect by considering only a small subset of pathways. For one set of parameters, path lengths up to six links in length had to be considered even to get the sign of the interaction right. The theoretical approach allows one to mathematically calculate all pathways. However, uncertainty in the estimates of the f_{ij} (*,*)s can lead to greater uncertainties in the calculation of long chains of indirect effects, as will be discussed further below.

There have been two recent surveys of experimental determinations of direct versus indirect effects by Schoener [25] and Menge [16]. The main conclusions are shown in Table 4.1. Schoener's survey showed that direct effects were stronger in six of the eight studies he surveyed, though indirect effects were more variable. Menge surveyed only studies of rocky intertidal food webs, 23 in all. He found that change in a given species abundance due to indirect effects ranged from 20 to 80% of the total change measured, with a mean of 45%.

Burns *et al.* [26] have applied the theoretical approach to comparing direct

Table 4.1 Empirical results concerning the relative magnitudes of direct and indirect effects. The authors conducted surveys of published field studies in which manipulations were performed to evaluate the relative importance of direct and indirect effects on given species populations

---

Schoener's (1993) survey of six experimental studies
  Direct effects stronger in 75% of experiments
  Indirect effects more variable in most experiments

Menge's (1994) survey of 23 marine intertidal experimental studies
  Identified 565 indirect interactions, classified into nine different types
  Of the 565 indirect interactions:
    34.7% were keystone predation
    25.0% were apparent predation
  while the others ranged from 7.4 to 2.8%
  Changes in density of a population due to indirect effects ranged from 20 to 80% of
    the total change and averaged 45%

---

and indirect effects in a lake ecosystem. The model used was the Standard Water Column Model (SWACOM) developed by Bartell *et al.* [27]. This model contains ten phytoplankton, five zooplankton, three planktivorous forage fish and a single piscivorous game fish in a column of lake water. It is possible to describe the dynamics of this system abstractly by the set of equations:

$$dX_i/dt = g_i(X_1, X_2, ..., X_n)\ (i = 1, 2, ..., n)$$

where X_i ($i = 1, 2, ..., n$) are population sizes and g_i are non-linear functions of X_i. Mathematically, the net total (direct and indirect) effects of sufficiently small changes in one species on another can be obtained by mathematical operations. In particular, a matrix of direct effects, $A = [a_{ij}]$, where $a_{ij} = \partial g_i/\partial X_j$, is first computed. The matrix $E = SB$ is then computed, where $S = -A^{-1}$ and B is the diagonal matrix with elements $-a_{jj}$. The elements e_{ij} of matrix E are then the total direct and indirect effects on X_i that result from a change in X_j. A description of the use of matrix methods used in describing the propagation of effects in food webs can also be found in [22].

Burns *et al.* [26] compared the direct and indirect effects on the game fish resulting from a unit perturbation to one of its forage fish species. They found that the net indirect effects were four times as great as the net direct effect.

4.5 FACTORS AFFECTING THE PROPAGATION OF INDIRECT EFFECTS

It is quite possible that idealized model studies may overestimate the impact of indirect effects. Schoener [25] pointed out that externally imposed stochasticity in species population abundances may decrease the importance of long pathways relative to short ones. To transmit a change in the manner predicted by the theoretical model, each species must respond in deterministic fashion. However, stochastic factors, such as environmental variations, can have their own effects on populations. For example, Spiller and Schoener [28], discussing food webs in Caribbean islands, noted that tropical storms can have a strong negative effect on spider numbers. If the stochastic effects are large compared to the deterministic signal, they can drown out the indirect effect, just as static can drown out a radio signal. This is more apparent the longer the pathway and the longer the time period that it takes the signal to travel from one species to another.

Another factor that can decrease the strength of an indirect effect is the presence of density-dependent 'self regulation' in species. This effect can arise through various negative effects that the size of a population has on its own growth rate. Suppose, for example, that a population of birds is limited by the number of available nesting sites. If there is an absolute limit on available nesting sites in an area, this can constitute a form of density-dependent regulation on the population, because the larger the bird population (if its adult numbers exceed the available sites) the smaller the per capita reproductive rate. In this case, a change in a species directly affecting the birds, such as a prey species, may not have an appreciable affect on the population size, which is already regulated by the availability of nesting sites. Therefore, a change in the abundance of an insect prey species will not be transmitted along an indirect pathway involving the bird species, or at least not as a very strong signal.

4.6 INDETERMINACY OF ECOLOGICAL INTERACTIONS

It is of practical importance to know whether the indirect effects of a change in a food web can be predicted or at least estimated. In particular, if a given species, i, is changed by a certain amount, can the effects, both direct and indirect, on another species, j, be calculated? The example of the approach of Burns *et al.* [26] described above indicates that calculations can be made in principle. However, these calculations were performed on model systems in which the parameters for the direct interaction strengths were assumed known. Actually, the interaction strengths between the species in a food web can only be estimated. Given that there is large uncertainty in these direct interactions, what will the uncertainty in the estimates of the indirect interactions be?

Yodzis [23] has examined this problem. His approach was to take 16 moderately complex food webs, which had been previously diagrammed, such as that for Narragansett Bay, Rhode Island (Figure 4.6), and assign an interaction matrix **A** with interaction strengths a_{ij}. Information on the interaction strengths was lacking, but Yodzis created a 'plausible community matrix', choosing these values randomly, such that each could range over an order of magnitude, but satisfying certain criteria of realism. Yodzis then applied 'press perturbations' to the web by adding a constant input to one of the species.

Table 4.2 Some results on the predictability of directionality in Yodzis's 1988 study [21]. The directionality of change of a species population due to the aggregate of direct and indirect effects from other populations was considered indeterminate if for 95% or more of the sample parameter sets the direction of change of a given species was the same

Type of effect	No. of effects	Proportion directionally undetermined	Fraction showing opposite of expected effect
Self	223	0.27	0
Predator on prey	317	0.52	0.11
Prey on predator	317	0.54	0.07
Indirect	3423	0.5	0.29
Competitive	566	0.58	–

One of the questions Yodzis then asked was whether the directional of change of species *i* could be predicted, given a unit change in species *j*. He termed the effect of species *j* on species *i* as directionally determined if the confidence level, given a set of community matrices sampled from the above ranges, was at least 95%. The results are shown in Table 4.2. For example, 'self' means the effect that a unit input has on the species itself after all indirect effects have been included. There are 223 species in the 16 webs. Yodzis found that in a fraction 0.27 of these it was not possible with 95% confidence to predict the direction of change of the species itself.

4.7 CONCLUSIONS

Indirect effects in ecology refer to effects of changes in one population on another population that involve changes in the population size or in the behaviours of other populations. Studies on certain target species of the effects of toxicants applied to ecosystems have revealed that indirect effects may be as important as direct effects of the toxicant on the target species [4]. Therefore, it is clear that ecotoxicology must take indirect effects into account. Fortunately, the study of indirect effects has become one of the central issues of ecology in recent years. How quickly this will lead to predictive capability is far from certain, however. Empirical studies in the field are difficult and expensive to conduct, particularly if these studies have proper controls and replicates. As a result, the study of indirect effects of toxicants may have to rely on mesocosm studies and model studies. Each of these has limitations. Models make simplifying assumptions and may omit important biological factors. Mesocosms are limited to small spatial scales. Furthermore, because mesocosms are often unstable over long periods of time, they may miss a majority of indirect effects that take long time periods to be manifested.

Since it is best to close on a positive note, it should be kept in mind that the increase in understanding of indirect effects in ecosystems has been rapid during the past few years. It is clear that much of the progress in understanding indirect effects is coming through basic research in ecology. However, as we have seen, ecotoxicological research has provided some striking examples of indirect effects in food webs. This is a further argument, if any is needed, for closer relationships between ecologists and ecotoxicologists.

REFERENCES

1. Anderson, D. W. and Hickey, J. J. (1972) Eggshell changes in certain North American birds. *Proceedings of an International Ornithological Congress*, **15**, 514–40.
2. O'Connor, R. J. and Shrubb, M. (1986) *Farming and Birds*, Cambridge University Press, Cambridge, UK.
3. Potts, G. R. (1981) Insecticide sprays and the survival of partridge chicks. *Game Conservancy Annual Review*, **12**, 39–48.
4. Talmage, S. S. and Walton, B. T. (1991) Small mammals as monitors of environmental contaminants. *Reviews of Environmental Contamination and Toxicology*, **119**, 47–145.
5. Barrett, G. W. and Darnell, R. W. (1967) Effects of dimethoate on small animal populations. *The American Midland Naturalist*, **77**, 165–75.
6. Hall, A. T., Woods, P. E. and Barrett, G. W. (1991) Population dynamics of the meadow vole (*Microtus pennsylvanicus*) in nutrient-enriched old-field communities. *Journal of Mammalogy*, **72**, 332–42.
7. Paine, R. T. (1966) Food web complexity and species diversity. *The American Naturalist*, **100**, 65–75.
8. Paine, R. T. (1969) A note on trophic complexity and community stability. *The American Naturalist*, **103**, 91–3.
9. Lubchenco, J. (1978) Plant species diversity in a marine intertidal community: Importance of herbivore food preferences and algal competitive abilities. *The American Naturalist*, **112**, 23–39.
10. Abrams, P. A., Menge, B. A., Mittelbach, G. G., Spiller, D. and Yodzis, P. (1996) The role of indirect effects in food webs, in *Food Webs: Integration of Pattern and Dynamics* (eds G.A. Polis and K.O. Winemiller), Chapman & Hall, Inc., New York, pp. 371–95.
11. Holt, R. D. (1977) Predation, apparent competition, and the structure of prey communities. *Theoretical Population Biology*, **12**, 197–229.
12. Holt, R. D. (1984) Spatial heterogeneity, indirect interactions, and the coexistence of prey species. *The American Naturalist*, **124**, 377–406.
13. Carpenter, S. R., Kitchell, J. F. and Hodgson, J. R. (1985) Cascading trophic interaction and lake ecosystem productivity. *BioScience*, **35**, 635–9.
14. Brown, J. H., Davidson, D. W., Munger, J. C. and Inouye, R. S. (1986) Experimental community ecology: The desert granivore system, in *Community Ecology* (eds J. Diamond and T. J. Case), Harper & Row, Publishers, New York, pp. 41–61.
15. Lima, S. L. and Dill, L. M. (1990) Behavioral decisions made under the risk of predation: A review and prospectus. *Canadian Journal of Zoology*, **68**, 619–40.
16. Menge, B. A. (1995) Indirect effects in marine rocky intertidal interaction webs: Patterns and importance. *Ecological Monographs*, **65**, 21–74.
17. Schindler, D. W. (1987) Detecting ecosystem responses to anthropogenic stress. *Canadian Journal of Fisheries and Aquatic Sciences*, **44**, 6–25.

18. Koivisto, S. (1992) *Model Ecosystems as a Tool in Aquatic Ecotoxicology*, Institutionen for Systemekologi, Stockholms Universitet.
19. Clements, W. H., Cherry, D. S. and Cairns, J. Jr (1989) The influence of copper exposure on predator–prey interactions in aquatic insect communities. *Freshwater Biology*, **21**, 483–8.
20. Borgmann, U., Millard, E. S. and Charlton, C. C. (1989) Effect of cadmium on a stable, large volume, laboratory ecosystem containing *Daphnia* and phytoplankton. *Canadian Journal of Fisheries and Aquatic Sciences*, **46**, 399–405.
21. Taub, F. B., Rose, K. A., Swartzman, G. L. and Taub, J. H. Translating population toxicity to community effects, Unpublished work.
22. Bender, E. A., Case, T. J. and Gilpin, M. E. (1984) Perturbation experiments in community ecology: Theory and practice. *Ecology*, **65**, 1–13.
23. Yodzis, P. (1988) The indeterminacy of ecological interactions as perceived through perturbation experiments. *Ecology*, **69**, 508–15.
24. Yodzis, P. (1996) Food webs and perturbation experiments: Theory and practice, in *Food Webs: Integration of Pattern and Dynamics* (eds G.A. Polis and K.O. Winemiller), Chapman & Hall, Inc., New York, pp. 192–200.
25. Schoener, T. W. (1993) On the relative importance of direct versus indirect effects in ecological communities, in *Mutualism and Community Organization* (eds H. Kawanabe, J. E. Cohen and K. Iwasaki), Oxford University Press, Oxford, UK, pp. 365–411.
26. Burns, T. P., Brenkert, A. L. and Rose, K. A. The measurement of indirect and direct effects in model ecosystems I. The steady-state case, Unpublished work.
27. Bartell, S. M., Brenkert, A. L., O'Neill, R. V. and Gardner, R. H. (1988) Temporal variation in regulation of production in a pelagic food web model, in *Complex Interactions in Lake Communities* (ed. Carpenter, S. R.), Springer-Verlag, New York, pp. 101–18.
28. Spiller, D. A. and Schoener, T. W. (1988) An experimental study of the effects of lizards on web-spider communities. *Ecological Monographs*, **58**, 57–77.

5 The dimensions of space and time in the assessment of ecotoxicological risks

PAUL C. JEPSON AND TOM N. SHERRATT

Incorporation of ecological concepts within ecotoxicological risk assessment procedures has been delayed partly because of a lag in the availability of appropriate ecological theory and partly because of the technical difficulty of penetrating procedures that are now well established and built into legal frameworks of pollutant regulation and control. This chapter explores, using simple modelling and case examples, the form that the case for taking more note of ecological arguments would have to take if general revisions to the procedures of ecotoxicological risk assessment were to be undertaken. It provides clear evidence of the potential benefits through reductions in the numbers of erratic risk assessments and improvements in our ability to manage chemical pollutants. An alteration in the form that risk assessment takes towards a more probabilistic evaluation of the likelihood that adverse effects may occur at the population level is, however, needed before ecological insights are more widely accepted.

5.1 INTRODUCTION

Ecotoxicology lies at the interface between the disciplines of toxicology, environmental chemistry and ecology. Arguably, the solution of all ecotoxicological problems requires insights from each discipline. Toxicology and environmental chemistry have, however, dominated in the development of ecotoxicological risk assessment and the role of ecology is still uncertain. Risk assessment procedures make little use of ecological concepts and regula-

ECOtoxicology: Ecological Dimensions. Edited by D. J. Baird, L. Maltby, P. W. Greig-Smith and P. E. T. Douben. Published in 1996 by Chapman & Hall, London. ISBN hardback 0 412 75470 3 and paperback 0 412 75490 8.

tory decisions for pollutants are based largely upon assessments of toxicological impact at the individual organism level and of environmental fate of the chemical concerned [1].

Ecotoxicologists have not developed general methods for determining the likely outcome of toxic effects at the population level. Risk assessment procedures determine the likelihood that individuals will be exposed to harmful doses of a pollutant, rather than the likelihood that toxic effects will lead to population decline, extinction, increase or neutrality.

Why does ecology contribute so little to ecotoxicology? The most likely reason lies in the enormous diversity of ecotoxicological problems, particularly the wide variation in relative scaling of chemical persistence and distribution, the mobility and reproductive rates of affected organisms, and the stability and layout of habitats which the pollutant and organism co-occupy. The toxicity and fate of pollutants can be measured within laboratory procedures and predictions of toxic effects on individuals, in the short term, may be very accurate; ecological attributes of habitat and organisms affecting risks may, however, require both laboratory- and field-based investigations over much longer timescales and may be highly site-specific. The incorporation of the ecological dimension adds uncertainty to risk predictions, an unwelcome addition when the needs for repeatability and cost effectiveness drive the design of most risk assessment methodologies. Through these needs, ecotoxicologists have sought refuge in toxicology and environmental chemistry, and risk assessment procedures commonly involve extrapolation of findings from simplified laboratory test systems or small mesocosms to the field using toxicological not ecological currencies to quantify the hazard posed by a pollutant to a particular organism. Given the difficulty of validating risk predictions, this approach has become widely accepted and is not the subject of serious scientific challenge.

The underlying rationale of ecotoxicological risk assessment should, as in all scientific disciplines, be subjected to regular challenge for reasons of scientific rigour and because of the importance of well constructed risk assessment procedures for the maintenance of environmental quality. Tests of current practices should be constructed to seek evidence of errors [2]. If errors are detected, then ways of reducing the likelihood of these taking place need to be sought. The case for changes to risk assessment procedures, especially the incorporation of the ecological approach, should not be made arbitrarily. Any major alteration needs to be justified in terms of quantifiable improvements in the regulation and management of environmental pollution and the degree to which it makes corrections for any important sources of error in current practice.

This chapter explores dimensions of time and space in ecotoxicology. It uses arguments from first principles, ecological theory and case studies to develop some generalizations about the role of ecological insight at all stages in the risk assessment process. By revealing scope for both falsely negative and falsely positive risk predictions within current systems, it makes the case for changes in practice and suggests ways in which ecological arguments may ameliorate these problems.

5.2 SPACE AND TIME IN ECOTOXICOLOGY: ARGUMENTS FROM FIRST PRINCIPLES

Reduced to their simplest mathematical form, all ecotoxicological scenarios must incorporate sufficient features of chemical fate and availability to be able to quantify the concentration of toxicant, sufficient attributes of habitat for spatial scaling of pollutant and organism distribution, and sufficient details of the organism including susceptibility (from which the effect of a given toxin concentration may be calculated), mobility (from which the likelihood of encountering a given toxicant concentration may be determined) and increase rate (from which population trajectories may be inferred) to enable the course of population density to be determined. An appropriate timescale, over which chemical and population processes evolve, must be established to generate the final outcome.

A simple mathematical approach which encapsulates these basic details may be expressed as follows.

1. A uniform area of habitat contains a proportion P, which is contaminated by a chemical pollutant. Area P is inhabited by x individuals of a particular species and the uncontaminated part of the habitat contains y individuals of the same species. All organisms, $x + y$, constitute a closed population, with free interchange between the unpolluted and polluted zones, defined by a diffusion rate, a.
2. Area P imposes a high rate of mortality, Z, caused by the pollutant once an organism enters this zone. Z declines over time according to a decay constant, q. Thus, the overall proportion of the population killed per unit time (t) is Ze^{-qt}.
3. The rate equations governing population dynamics in the contaminated zone (P) and the uncontaminated zone are as follows:

   (a) in the contaminated habitat: $dx/dt = rx(1 - x/PK) - ax + ay - xZe^{-qt}$
   (b) in the uncontaminated habitat: $dy/dt = ry[1 - y/(1 - P)K] + ax - ay$

   where r is the intrinsic increase rate and K is the carrying capacity of the habitat.

In the worst case scenario, where the pollutant in area P is completely toxic (i.e. $Z = 1$) and permanent, then the organism can still persist in the system, as long as $r > a$. In other words, the rate of contributing new individuals to the population must exceed the rate at which they enter the polluted part of the habitat and die.

Several numerical solutions to the rate equations are shown in Figure

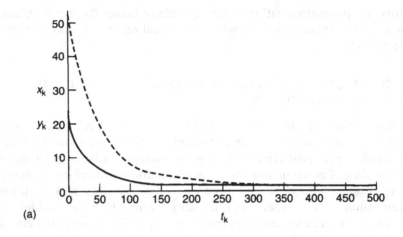

(a)

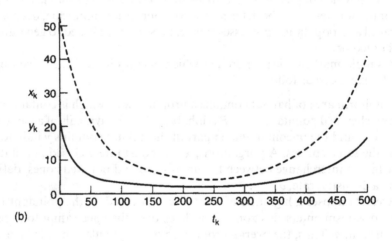

(b)

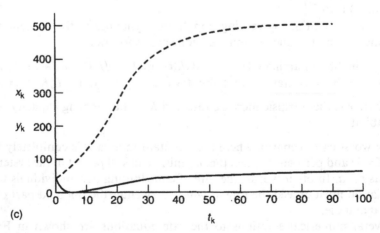

(c)

5.1a–c. The results are rather predictable and intuitive in such a simple system; they do however reveal the interplay between key parameters and the potential for small changes to lead to markedly differing outcomes. In the example in Figure 5.1a, high mortality, with infinite chemical persistence, leads to global extinction because the source area becomes exhausted. In this example, increase rate within the source area cannot compensate for losses within the polluted zone and the only effect of having part of the habitat unpolluted is to delay the eventual demise of the population. In the example illustrated in Figure 5.1b, a very slow decay of pollutant effectively rescues an otherwise identical system from extinction. There are delicate trade-offs between the rate of diffusion, chemical persistence and reproductive rate, such that extinction may occur if, for example, the likelihood of entering the polluted zone is increased. This relationship is explored in Figure 5.1c, where the diffusion rate is reduced, yielding a new set of equilibria and the maintenance of high density in the source area.

Chemical toxicity and organism susceptibility were held constant throughout these examples and the potential for exposure to the pollutant did not vary within the polluted zone. Risk assessment relying upon toxicity and environmental concentration measurements alone would have made identical predictions in each of the scenarios explored in Figure 5.1. The actual outcome varied, however, from extinction to densities approaching the carrying capacity, as a function of movement and reproductive rates, in a complex interplay with the decay rate of the pollutant.

What is the significance of this finding for risk assessment? Firstly, the analysis re-emphasizes the importance of taking into account the relative scaling of temporal and spatial processes as they affect the chemical (i.e. distribution and decay rate), the habitat (i.e. heterogeneity, fragmentation with respect to distribution, and movement of chemical and organism) and the organism (i.e. movement and increase rates). It is apparent that the eventual outcome of chemical exposure (which ought to provide us with our currency of risk) is critically dependent upon all of these. Secondly, the analysis illustrates scope for generating errors in risk assessment procedures that fail to take into account ecological processes. Falsely positive results, i.e. predictions of a hazardous outcome that could not arise in the real world, may indicate that the method of calculating risks is appropriately conservative. This result is, however, still in error. Most worrying must be the potential for falsely negative results, i.e. predictions of acceptable impacts when the

---

Figure 5.1 *(facing)* Population trajectories for organisms within polluted, x (—), or unpolluted, y (- - -), zones of a closed habitat (for details see text). (a) High toxicity ($Z = 1.0$), very high persistence (decay, $q = 0$), low diffusion rate (exchange, $a = 0.3$). (b) High toxicity ($Z = 1.0$), high persistence (decay, $q = 0.001$), low diffusion rate (exchange, $a = 0.3$). (c) High toxicity ($Z = 1.0$), high persistence (decay, $q = 0.001$), very low diffusion rate (exchange, $a = 0.1$).

outcome is hazardous in the real world. Even compounds with low toxicity in the simple model above may prove hazardous and cause global extinction if the pollutant is widespread, or the organism mobile enough that exposure is likely and reproduction fails to replace those organisms that succumb to toxic effects.

Is there evidence, therefore, that temporal and spatial processes trade-off against each other in the real world in such a way that these errors could arise?

5.3 EFFECTS OF TEMPORAL AND SPATIAL FACTORS ON ECOTOXICOLOGICAL PROCESSES

5.3.1 EXAMPLE 1: LINYPHIIDAE IN SPRAYED FARMING SYSTEMS

(a) Risk assessment based upon toxicity and exposure

Linyphiidae are very susceptible to synthetic organic pesticides in both laboratory- and field-based testing, and both spiders and their webs are highly exposed to direct spraying and to toxic residues [3, 4]. Using conventional methods of risk assessment from laboratory- through to field-based testing [5], high risks are predicted in terms of the probability of death in the period following spray application.

(b) Spatial and temporal processes mediating risk

Linyphiidae are highly dispersive and pesticides have short half-lives, such that repopulation of sprayed areas is likely under most circumstances (e.g. [6, 7]). Spiders complete their life cycles in more than one field however and the complex interplay between habitat, chemical and organismal factors that operate on scales that are appropriate to linyphiid population dynamics may only be explored using mathematical models [8]. These reveal that high risks from spraying occur when spray application is frequent and widespread; these risks are, however, dependent upon the degree of dispersal by spiders, the availability of favourable habitat beyond sprayed fields and the degree of synchrony in spraying between fields.

(c) Conclusions

Only by incorporating habitat and organismal parameters within models is it possible to define those situations where short-term toxicological risk may be translated into high long-term risk. Conventional risk predictions, based upon toxicological criteria, appear falsely positive when viewed from an ecological perspective. The scope for falsely negative predictions in the case of such dispersive organisms is limited [2, 8]: products of low toxicity and persistence are unlikely to present a problem even when widely applied. Linyphiidae appear to be sensitive to changes in the nature of the larger scale ecosystem and pattern of landscape management over the large area where their life cycles are completed. Pesticide impact upon Linyphiidae in a more uniform system (e.g. wholly sprayed cereals) or a system that contained fewer alternative habitats could be very different.

(d) Support from theoretical ecology

Linyphiidae represent the type of 'patchy population' [9, 10] that is particularly sensitive to factors that alter the nature of the whole system. They are so dispersive that populations in individual patches are effectively synchronized with each other. The break up of agricultural land by field boundaries and crops that present differing degrees of hazards to spiders, structures the population such that it resembles a metapopulation. Linyphiid populations are, however, unlike classical 'metapopulations', where populations in isolated patches exhibit a degree of asynchrony in their dynamics and where the persistence of the metapopulation is a function of interpatch dynamics [11]. The persistence of Linyphiidae is less likely to be controlled by the balance between colonization and extinction rates in individual patches because patches are effectively synchronized and subject to the same external influences. The focus of research and the measurements made to assist the interpretation of risk assessment should therefore be on a larger scale than might be the case for less dispersive organisms.

5.3.2   EXAMPLE 2: LOCUST HOPPER BANDS EXPOSED TO PERSISTENT PESTICIDES

(a) Risk assessment based upon toxicity and exposure

Pesticides used against locusts are potentially damaging to fragile Sahelian environments [12]. Most classes of conventional pesticide are highly toxic to both locusts and non-target species, and risk assessment predicts, for most products, severe effects on many invertebrate taxa [13]. On the basis of risk assessment alone, there would not appear to be any scope for partitioning these hazardous effects in favour of locusts, such that non-target species are less affected.

(b) Spatial and temporal processes mediating risk

Following the arguments within this chapter, however, an ecological perspective of the spatial and temporal factors mediating risk indicates some scope for selective use of highly persistent chemicals. In theory, there will be trade-offs between the dispersal rate of the organism, the persistence of the pesticide and the proportion of the habitat that is treated. If the target species was one of the most mobile invertebrate taxa in the sprayed environment then, as chemical persistence increases, for a given level of intrinsic toxicity, the same effect could be achieved with increasingly smaller proportions of the habitat sprayed, with consequent benefits for less dispersive, non-target, taxa. This prediction has been confirmed in practice. Highly persistent chemicals may be sprayed in well separated barriers [14, 15] and still achieve a very effective kill of locusts. Non-target species, which appear less dispersive, are not significantly affected in the unsprayed zone (Colin Tingle, Natural Resources Institute, Chatham, UK, personal communication).

(c) Conclusions

Only by appreciating the trade-off between persistence, movement and effects was it possible to devise the barrier spray application tactic with its real environmental benefits. Conventional risk assessment would have placed all chemicals in a similar risk category and would fail to arrive at appropriate methods of chemical management. Although this approach to control was developed to maximize kill of locusts per unit logistical effort, with little thought for environmental benefits, there are lessons for ecotoxicologists. By excluding spatial and temporal considerations from risk assessment, ecotoxicologists miss opportunities for effective extrapolation of their test results to the real world. They also increase the likelihood of error (falsely positive predictions in the case of non-dispersive invertebrates colonizing the areas between sprayed barriers) and limit the potential for effective pollution management tactics arising from their analyses.

(d) Support from theoretical ecology

In this case, although generally supporting the line of argument in this chapter overall, there is little scope for support from theoretical ecology because the chemical impact is targeted at one life-cycle stage of the locust, i.e. the wingless, marching nymph, and is not logically extendible to consider population processes between generations. It is interesting to note that barrier spraying was developed in the 1940s, when persistent organochlorine insecticides were first available, and were the product of practical experience not application of theory. Herein lies another message for ecotoxicologists – that monitoring and validation are needed in the real world in order to gain an appreciation of the importance of habitat and organismal factors mediating risk. Much too can be learnt by determining the relative scaling of the different processes that mediate risk in the real world. Applied entomologists and pesticide specialists seem to be more advanced than ecotoxicologists in recognizing the importance of dispersal as a mediator of chemical effects.

5.3.3 EXAMPLE 3: GROUND BEETLES IN SPRAYED FARM SYSTEMS

(a) Risk assessment based upon toxicity and exposure

Carabidae are medium to large beetles with moderate to low susceptibility to pesticides [16, 17]. They are conventionally used in laboratory-based risk assessment test procedures [5] because they represent an important guild of natural enemies.

(b) Spatial and temporal processes mediating risk

Although risk assessment procedures would not necessarily rank Carabidae highly, compared to other natural enemies, including Linyphiidae, this family appears particularly susceptible to severe long-term effects and even local extinction [18]. Risk assessment is again a poor guide to the eventual outcome because it fails to consider the relative scaling of spatial and temporal factors that mediate effects in the real world. In this case, Carabidae, being poor dispersers, including many wingless species with low rates of reproduction, are particularly susceptible to increasing spray frequency and to factors, such as dispersal rate and field boundary permeability, that control colonization rate [19]. Carabidae appear so sensitive to factors that have small effects on survival and colonization rates that even mildly toxic insecticides could cause local extinctions if used frequently enough, or on a large enough scale.

(c) Conclusions

Such false negatives arising from risk assessment could permit continued use of insecticides in situations with a high probability of adverse long-term effects. The lessons for ecotoxicological risk assessment include further reinforcement of the need to incorporate ecological considerations at an early stage in any system where risk is scale-dependent.

(d) Support from theoretical ecology

Populations of Carabidae, separated by field boundaries, represent classical metapopulations [11]: arable landscapes are sub-divided into fields and hazardous farming practices are distributed asynchronously between these in such a way that populations probably persist through a balance between recolonization and extinction dynamics. Insight from theory could contribute a great deal to the design, analysis and interpretation of ecotoxicological risk assessment associated with true metapopulations, exposed to patchily distributed pollutants [2]. It is not thought, however, that classical metapopulation structures are widely distributed in nature [9, 10] and metapopulation theory may not have the potential for being widely taken up within ecotoxicological risk assessment.

5.4   THE NEXT STEPS IN ECOTOXICOLOGY

5.4.1   'RULES OF THUMB' AND THE INCORPORATION OF ECOLOGICAL THEORY

A practical and elegantly argued procedure for exploiting spatial dynamics and landscape ecology, to be followed by ecotoxicologists, has been outlined by Fahrig and Freemark [20]. They suggest that simple models should be developed, incorporating the spatio-temporal dynamics of populations and toxic events, and that predictions from these should be validated by field observation.

Although the carabid and linyphiid case examples above provide general predictions about pesticide risks from models, neither incorporates the structure of real landscapes nor is sufficiently species-specific to be eligible as a predictive tool for individual risk assessments. This represents the main limitation to this approach: the high level of detail needed in order to be able to correctly assign risk in a real system. Field experience tells us that Carabidae in sprayed farm systems may exhibit extinction, superabundance or neutrality in the face of heavy spraying regimes [18]. Although the processes underlying population dynamics and responses to spraying are likely to be explained by models such as those developed by Sherratt and Jepson [19], the biological, chemical and toxicological detail required to be able to predict exactly which population trajectory a given species will follow makes it unlikely that models alone will suffice for risk assessment.

Ecotoxicologists must ask more often 'What level of detail in understanding is required to explain the outcome of a given toxic episode?'. Only by building experience can the appropriate trade-off between model and experiment be struck. Ecological theory has made enormous progress in its ability to explain trends in population persistence and extinction. The ecologist will always recognize, however, that there are many possible outcomes from a given set of starting conditions. The hard and reliable predictions of laboratory-based risk assessment do not readily fit with this view of the world; they can however be placed in an ecological context and greatly enrich the value of the data obtained. The cost of this approach is an apparent loss of precision. The gain may be that we will get closer to predicting the outcome that is most likely. Most importantly, if ecotoxicologists accept that current methodology carries an unpredictable risk of false negatives and false positives, it can be argued that these risks are minimized by the incorporation of more detail of the spatio-temporal dynamics of pollutant, habitat and organism.

REFERENCES

1. Calow, P. (ed.) (1993) *Handbook of Ecotoxicology, Vol. 1*, Blackwell Scientific Publications, Oxford, UK.

2. Jepson, P. C. (in press) Scale dependency in the ecological risks posed by pollutants: is there a role for ecological theory in risk assessment?, in *Ecological Principles for Risk Assessment of Contaminants in Soil* (eds N. Van Straalen and H. Lokke), Chapman & Hall, London, UK.

3. Everts, J. W., Willemsen, M., Stulp, M., Simons, B., Aukema, B. and Kammenga, J. (1991) The toxic effect of deltamethrin on linyphiid and erigonid spiders in connection with ambient temperature, humidity and predation. *Archives of Environmental Contamination and Toxicology*, **20**, 20–4.

4. Stark, J. D., Jepson, P. C. and Thomas, C. F. G. (1995) The effects of pesticides on spiders from the lab. to the landscape. *Review of Pesticide Toxicology*, **3**, 83–110.

5. Jepson, P. C. (1993) Insects, spiders and mites, in *Handbook of Ecotoxicology, Vol. 1* (ed. P. Calow), Blackwell Scientific Publications, Oxford, UK, pp. 299–325.

6. Thomas, C. F. G., Hol, E. H. A. and Everts, J. W. (1990) Modelling the diffusion component of dispersal during the recovery of a population of linyphiid spiders from exposure to an insecticide. *Functional Ecology*, **4**, 357–68.

7. Thacker, J. R. M. and Jepson, P. C. (1993) Pesticide risk assessment and non-target invertebrates: integrating population depletion, population recovery and experimental design. *Bulletin of Environmental Contamination and Toxicology*, **51**, 523–31.

8. Halley, J. M., Thomas, C. F. G. and Jepson, P. C. (in press) A model of the spatial dynamics of linyphiid spiders in farmland. *Journal of Applied Ecology*.

9. Harrison, S. (1991) Local extinction in a metapopulation context: an empirical evaluation, in *Metapopulation Dynamics: Empirical and Theoretical Investigations* (eds M. E. Gilpin and I. Hanski),. Academic Press, London, pp. 72–88.

10. Harrison, S. (1994) Metapopulations and conservation, in *Large Scale Ecology and Conservation Biology* (eds P. J. Edwards, R. M. May and N. R. Webb), Blackwell Scientific Publications, London, pp. 111–29.

11. Gilpin, M. E. and Hanski, I. (eds) (1991) *Metapopulation Dynamics: Empirical and Theoretical Investigations*, Academic Press, London.

12. Everts, J. W. (ed.) (1990) Environmental effects of chemical locust and grasshopper control. Project report ECLO/SEN/003/NET, FAO, Rome.

13. Murphy, C. F., Jepson, P. C. and Croft, B. A. (1994) Database analysis of the toxicity of antilocust pesticides to non-target, beneficial invertebrates. *Crop Protection*, **13**, 413–20.

14. Bouaichi, A., Coppen, G. D. A. and Jepson, P. C. (1994) Barrier spray treatment with diflubenzuron (ULV) against gregarious hopper bands of the Moroccan Locust (*Dociostaurus maroccanus* Tunberg) (Orthoptera: Acrididae) in N. E. Morocco. *Crop Protection*, **13**, 60–72.

15. Cooper, J. F., Coppen, G. D. A., Dobson, H. M., Rakotonandrasana, A. and Scherer, R. (1995) Sprayed barriers of diflubenzuron (ULV) as a control technique against marching hopper bands of migratory locust *Locusta migratoria capito*. (Sauss.) (Orthoptera: Acrididae) in Southern Madagascar. *Crop Protection*, **14**, 137–43.

16. Wiles, J. A. and Jepson, P. C. (1992) The susceptibility of a cereal aphid pest and its natural enemies to deltamethrin. *Pesticide Science*, **36**, 263–72.

17. Jepson, P. C., Efe, E. and Wiles, J. A. (1995) The toxicity of dimethoate to predatory Coleoptera: developing an approach to risk analysis for broad spectrum pesticides. *Archives of Environmental Contamination and Toxicology*, **28**, 500–7.

18. Burn, A. J. (1989) Interactions between cereal pests and their predators and parasites, in *Pesticides and the Environment: the Boxworth study* (eds P. Greig-Smith, G. E. Frampton and A. Hardy), HMSO, London, pp. 110–31.

19. Sherratt,T. N. and Jepson, P. C. (1993) A metapopulation approach to modelling the long-term impact of pesticides on invertebrates. *Journal of Applied Ecology*, **30**, 696–700.
20. Fahrig, L. and Freemark, K. (in press) Landscape scale effects of toxic events for ecological risk assessment, in *Toxicity Testing at Different Levels of Biological Organization* (eds J. Cairns and B. Niederlehner), Lewis Publishers, Boca Raton, FL.

6

Coping with variability in environmental impact assessment

JOHN A. WIENS

Although the responses of people to environmental accidents are often subjective and emotional, scientific evaluation of environmental impacts must be
founded on careful study and rigorous analysis. Assessments of environmental
impacts are often based on the presumption of a 'balance of nature'. Environments vary in time and space, however, and this invalidates the assumption of
equilibrium and complicates study design. Natural variability also influences
the likelihood that a study design will suffer from pseudoreplication due to
non-independence of samples, and it affects the power of statistical tests and
the effect size that can be documented. I evaluate the degree to which assumptions of temporal or spatial equilibrium are contained in various study designs
for assessing environmental impacts and determining recovery. Traditional
before–after or treatment-control designs rely on assumptions of a steady-state
equilibrium in time or space and may be of limited value. Other designs that
involve comparisons among paired sites or multiple-time sampling from areas
exposed to different perturbation intensities have greater capacity to separate
the effects of natural variation from those of the perturbation and may be less
sensitive to pseudoreplication. Coping with variability in environmental impact assessment requires careful attention to the assumptions inherent in study
designs and careful consideration of the relationships between variability,
pseudoreplication and power.

6.1 INTRODUCTION

Human activities inevitably affect organisms and the environment. These impacts occur through the harvesting of populations, the alteration or destruction

ECOtoxicology: Ecological Dimensions. Edited by D. J. Baird, L. Maltby, P. W. Greig-Smith
and P. E. T. Douben. Published in 1996 by Chapman & Hall, London. ISBN hardback
0 412 75470 3 and paperback 0 412 75490 8.

of habitats, or the release of contaminants into the environment. Although such activities affect the physical, chemical or biological components of environments, they occur within a cultural and socio-political context in which the environment has a 'value' that is based on many criteria: economics, aesthetics, recreational, religious and ecological, among others. As a result, the response of people to environmental alteration or contamination is often emotional and subjective, and these reactions may carry over into political action and policy formulation. To counteract such emotionalism, there is a special need to assess the environmental consequences of human activities using objective, rigorous, scientific approaches, either in predicting the potential effects of environmental alterations (ecological risk assessment) or in 'post-dicting' the consequences of environmental changes or accidents that have occurred (environmental impact assessment).

Designing studies to assess actual or potential environmental impacts and recovery from these impacts is complicated by both logistical and environmental constraints. Often, for example, environmental alterations are accidental (e.g. processing plant malfunctions, oil spills, forest fires, releases of toxic wastes from mine spoils). Such accidents are by definition unplanned and their occurrence is not random in space or time. Consequently, traditional research designs based on randomization of treatments and suitable replication are usually precluded [1]. A more fundamental constraint on the design of studies, however, arises from the natural variability of environments.

My objective here is to consider how we can design scientific studies to assess environmental risks and impacts in a variable world. Because environmental accidents are the most dramatic (although not necessarily the most important) forms of environmental contamination and pose the greatest challenges to study design, I will focus my comments on detecting the impacts of and recovery from such events. My conclusions, however, should apply more broadly, to the design of studies to assess the consequences of all sorts of environmental changes.

6.2 THE NATURE OF ENVIRONMENTAL VARIATION

The view that natural systems are in equilibrium has long been a dominant theme in western thought ('the balance of nature'; [2, 3]). This view has also permeated efforts to model ecological systems [4]. The vast proliferation of ecological theory based on linear difference and differential equations, in which equilibrium solutions have been the target of interest, has reinforced the basic belief that natural systems should also seek these solutions.

'Equilibrium', of course, may be defined in various ways [4, 5]. Ecological theory has generally considered either steady-state equilibrium, in which the state of a system remains unchanged over time or space (there is a constant mean about which numbers vary in a regular way), or dynamic equilibrium, in which various features of a system change in the same way over time in different locations, usually in response to broad-scale, regional control (e.g. climate).

Although ecological systems are often persistent (e.g. populations continue to be present in an area over a long time span), they generally are not in numerical equilibrium in the sense portrayed by mathematical or physical theory. Variability, not equilibrium, is the normal state of nature.

Examples abound. In a salt marsh in The Netherlands, densities of the plant *Salicornia brachystachys* varied 12-fold over a 13 year period, and spatial variation over a 21 m transect was even greater [6]. Annual recruitment of plaice (*Pleuronectes platessa*) in the North Sea varied from c. 200×10^6 to nearly 1200×10^6 over a 40 year period, with peak abundances often (but not always) coinciding with cold winters [7]. In Britain, abundances of redstarts (*Phoenicurus phoenicurus*) declined sharply from a peak in 1965 to low numbers in the early 1970s and then increased, returning to previous levels in some areas but not in others. Abundances of grey partridge (*Perdix perdix*), on the other hand, have declined more or less continuously since the First World War. The drop in redstart abundance has been attributed to prolonged drought on the wintering grounds in the Sahel of Africa, while the decline of partridges may have been due to the combined effects of changes in agricultural cropping practices, increased use of pesticides and a reduction in the number of game-keepers (and an attendant increase in predation on nests) [8]. Spatial variation in population density and dispersion is a well-known phenomenon for plant populations [9, 10] and is the foundation of considerations of ecological heterogeneity [11, 12] and metapopulation theory [13]. Various distributional atlases (e.g. [14–17]) and survey reports (e.g. [18]) illustrate spatial variation in bird abundances at regional scales.

Variation in the components of ecological systems arises from three sources: changes in the physical (or chemical) environment, intrinsic changes in the biota (e.g. density-dependent population dynamics) and sampling or estimation error. Moreover, each of these sources of variation may contain both deterministic and stochastic components, and the relative importance of these components may change with the scale of analysis. The effects of environmental variations on population size or recruitment, for example, may often appear to be stochastic at very fine scales but deterministic at broader scales [19]. Stochastic factors may have fixed probabilities of occurrence per unit time, although these probabilities differ among factors. With longer time periods, a larger number of factors is likely to be involved in determining the ecological responses of interest and interactions among these probabilistic factors may erode any semblance of predictability (or stability) of patterns. Even if factors act deterministically, non-linearities and time lags may produce complex dynamics in which the state of the system is strongly influenced by initial conditions [20–22].

Perhaps because of the dominance of equilibrium thinking in ecology, all of these sources of variation are often regarded as 'noise' and are combined in the error term in statistical analysis. It is easy to see why the notion of equilibrium has been so appealing, especially in theoretical work. The use of equilibrium approaches, however, should be carefully circumscribed in both theoretical and (especially) applied work [3, 21], and attention should be focused on separating the components of variability rather than regarding them collectively as 'error' [23].

6.3   VARIATION AND IMPACT ASSESSMENT

In attempting to forecast or assess environmental impacts, the principal concern is usually to separate the effects of human perturbations from the various other sources of variation (what is usually termed 'natural variation') [1, 24–26]. This requires some means of defining or recognizing impacts and subsequent recovery. Despite the obvious importance of natural variation, the tradition in environmental impact assessment still appears to be based upon equilibrium thinking. Thus, in evaluating the effects of the *Exxon Valdez* oil spill on various natural and human resources, recovery has been defined as 'a return to prespill conditions', or, for resources that were in decline before the spill, 'recovery may consist of stabilizing the population at a lower level than before the spill' [27]. When information on prespill levels is not available, recovery occurs 'when differences between oiled and unoiled areas are eliminated' [27]. In the first case, recovery is explicitly defined in terms of attainment of a steady-state equilibrium in time; in the latter case, equilibrium among locations in space is assumed.

In a variable world, such definitions are unrealistic. Instead, 'impact' and 'recovery' should be defined in relation to the background of natural variation that characterizes a system, in terms that encourage statistical analyses. Elsewhere [28], I have defined the impact of a contaminant such as oil as a statistically significant difference between samples exposed to the contaminant and reference (or 'control') samples that encompass the normal patterns of variation in the system. Recovery is the disappearance through time of such a statistical difference (Figure 6.1).

6.4   STUDY DESIGNS AND THEIR ASSUMPTIONS

Against this background of natural variation, it is essential to plan carefully research aimed at determining whether an anthropogenic perturbation has affected the environment and, if so, whether recovery has occurred. How can we design studies to cope with this variability? The definitions of recovery from the *Exxon Valdez* Trustee Council, quoted above [27], immediately suggest two possibilities: before–after comparisons for specified locations or comparisons between contaminated (treatment) and uncontaminated (reference or control) sites. Because the definitions are linked with equilibrium thinking, however, it is possible that these designs may reflect the same sort of thinking. To what

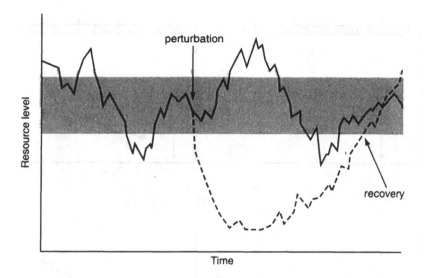

Figure 6.1 An hypothetical example of natural variability and environmental impact. The solid line traces the natural variation in levels of a resource over time; the shaded zone represents a 'window' of variability (e.g. 95% confidence interval). An anthropogenic perturbation may shift the resource from this natural trajectory (- - -). An 'impact' occurs when the perturbed system falls outside of the window of natural variation (as gauged by statistical tests). 'Recovery' then can be defined by the return of the perturbed system to the window of natural variability. Note that the resource level in the 'recovered' system does not necessarily match the precise value that might have characterized the system had the perturbation not occurred. After Wiens [28].

degree do these and other study designs contain assumptions about equilibrium in time or space? The following evaluation is based largely on that of Wiens and Parker [1], who give examples of each design from studies conducted following the *Exxon Valdez* oil spill. Other aspects of study design in environmental impact assessment have recently received considerable attention (e.g. [24, 25, 29–37]).

6.4.1 TIME-BASED DESIGNS

Some study designs are based on comparisons of the same sites (usually only a few) at different times. If measurements of environmental components of interest (hereafter referred to as 'resources') were made at a site before the environmental perturbation occurred, these values may be compared with measurements taken at one or more times after the perturbation (Figure 6.2a). This simple 'before–after' or **baseline** design is frequently used in assessing planned impacts. The expectation is that the values (or means of values) from

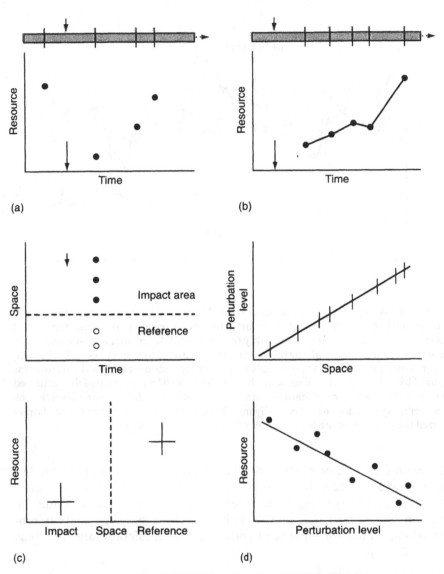

Figure 6.2 Diagrammatic examples of study designs used to assay environmental impacts and subsequent recovery. Each part of the figure (a–h) contains two diagrams: the top diagrams show the general structure of a particular study design, while the lower ones illustrate the sort of results that might indicate an impact of an environmental perturbation (shown where appropriate by the vertical arrows). (a) Baseline design: the upper shaded bar indicates a time block within which an accidental perturbation (vertical arrow) occurs; the vertical lines indicate samples taken before and after the perturbation. The lower graph depicts the levels of a resource at these sampling times. (b) Time series: like (a), except that no samples are available from before the

(Figure caption continued on page 62)

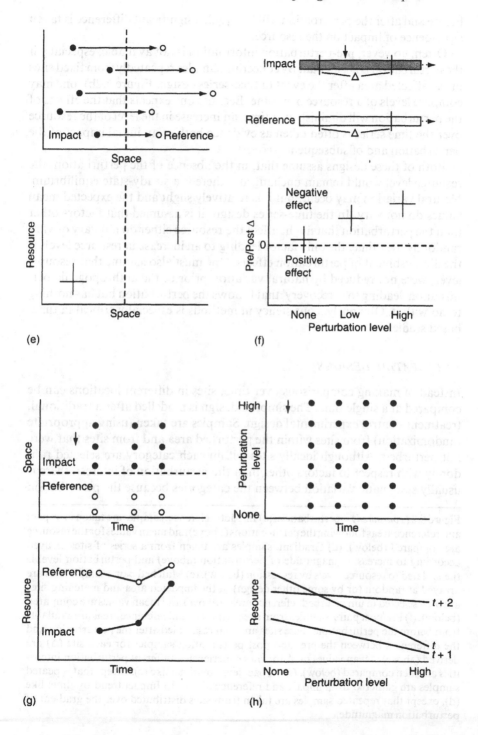

(e)

(f)

(g)

(h)

before and after the perturbation will be equal; a significant difference is taken as evidence of impact on the resource.

Often, however, pre-perturbation information is not available, especially if the perturbation is accidental. By collecting samples repeatedly from fixed sites in the affected area after the event (a **time-series** design; Figure 6.2b), one may compare levels of a resource over time. Because one expects that the effects of the perturbation will diminish over time, an increase in the level of the resource over the time series is often taken as evidence both of an initial impact of the perturbation and of subsequent recovery.

Both of these designs assume that, in the absence of the perturbation, the resource level would remain unchanged – there is a steady-state equilibrium. Natural variation may occur, but it is relatively slight and the expected mean values do not vary. In the time-series design, it is assumed that factors other than the perturbation that might affect the resource either do not vary or vary randomly over time; the only factor leading to an increase in resource levels is the diminishment of perturbation effects. One must also assume that resource levels were not reduced by natural variation prior to the anthropogenic per-turbation, leading to a 'recovery' that follows the perturbation but has nothing to do with it. Obviously, consistency in methods is especially critical in time-based studies.

6.4.2 SPATIAL DESIGNS

Instead of making comparisons over time, sites in different locations can be compared at a single time. The simplest design is modelled after a traditional, treatment-control experimental design. Samples are taken (using appropriate randomization) from sites within the perturbed area and from sites that were not perturbed. Although ideally sites within each category are selected ran-domly with respect to factors other than the perturbation of interest, there is usually systematic variation between the categories because the perturbation

---

Figure 6.2 (continued)   perturbation. (c) Impact–reference: samples are taken in impact and reference areas following the perturbations (above) and mean values for the resource are compared (below). (d) Gradient: samples are taken from a series of sites arrayed according to increasing magnitude of perturbation (above) and perturbation level is then related to resource levels by regression (below). (e) Matched pairs: sample sites are arrayed at random (or by some other design) in the impacted area and matching sites are then selected in unperturbed reference areas (above) and mean values are compared (below). (f) Pre/post pairs: samples from sites in impact and reference areas are available from before the perturbation; these sites are then resampled after the perturbation and the differences between the pre- and post-perturbation samples for each site (Δ) are derived (above). Mean values of Δ for sites experiencing different perturbation intens-ities are then compared (below). (g) Impact level-by-time: like (c), except that repeated samples are gathered from impact and reference areas. (h) Impact trend-by-time: like (d), except that repeated samples are taken from sites distributed over the gradient of perturbation magnitude.

itself occurs non-randomly. As a consequence, the unperturbed sites are not true controls, and it is more appropriate to refer to this as an **impact–reference** design (Figure 6.2c). If an impact has occurred, one expects that levels of the resource in the impact samples will differ systematically from those in the reference samples. Sometimes it is possible to categorize sites by the degree of perturbation (e.g. low, moderate or high disturbance), in which case levels of exposure may be used as treatments in an ANOVA design. If the degree of perturbation can be measured as a continuous rather than a discrete variable (e.g. a quantitative measure of contamination), then one may use a **gradient** design to sample from sites arrayed across the perturbation gradient (Figure 6.2d). The expectation is that the degree of injury to the resource should be related to the intensity of perturbation (i.e. a negative regression of resource level on perturbation magnitude).

In both of these designs, the investigator must assume that spatial variation is unimportant. That is, the effects of factors other than the perturbation on the resource are equal among sites and do not confound the relationship between the resource and the perturbation. One way to reduce the confounding effects of spatial variation is to measure other environmental factors at the sampling sites and use these as covariates in statistical analyses (e.g. [39]). Another way is to use a **matched pairs** study design. In a matched pairs design, sites within the perturbed area are randomly selected. Reference sites are then intentionally selected to match the environmental features (other than the perturbation) in the impact sites; each reference site is paired with a similar impact site (Figure 6.2e). Because the matched pairs are selected to be similar in factors that may affect the resource, one expects that any difference between them will be due to the perturbation. Although this design can control for spatial variation among sites, one must assume that the sites have been correctly paired (i.e. they do not differ in unmeasured factors that affect the resource). Moreover, because the reference sites are selected subjectively, there is an increased potential for conscious or unconscious bias to enter into the selection process.

6.4.3 TIME–SPACE DESIGNS

In the above designs, time and space are considered separately. As a result, when comparisons are made over time, spatial variation is assumed to be unimportant and spatial comparisons assume that temporal variation does not matter. Restricting spatial comparisons to a single time period, however, does not obviate the effects of natural variation in time. Because sites may differ in their recent histories, they may be at different phases in their trajectories of variation. Such differences increase variance (which has implications for statistical power; see below), but they may also produce systematic biases in the analyses if, for example, the recent history of impact sites differs in the same way from that of reference sites. One implicitly assumes temporal equilibrium (or, in temporal comparisons, spatial equilibrium).

Several study designs incorporate both temporal and spatial dimensions, and therefore make these assumptions more explicit (and testable). One is a variation on the baseline design. If information is available from several sites surveyed before the perturbation, these sites may be categorized by disturbance level and resampled after the event. In this **pre/post pairs** design the pre and the post samples are paired for each site and the analysis conducted on **differences** between the time periods (Figure 6.2f). Because the same site is resampled, the effects of spatial, among-site variation that might confound the effects of the perturbation are reduced considerably. The expectation is that mean pre/post differences should be equal if there is no injury to the resource but should correlate negatively with perturbation level if there is an effect. Although environmental factors may vary among the sites and over time, one assumes that the pattern of differences among sites at one time will be the same at a later time; there is 'temporal coherence' [24, 39]. Instead of assuming a steady-state equilibrium, this design assumes a dynamic equilibrium among sites.

Other time–space designs are variations on the impact–reference and gradient designs discussed earlier. If the basic impact–reference design is repeatedly sampled over time (an **impact level-by-time** design, Figure 6.2g), one can determine not only if there is a possible perturbation effect but whether these effects diminish over time. Because of natural variation, the resource level may change over time. A difference in the magnitude of these changes between impact and reference samples, however, suggests injury to the resource; when the resource has recovered, the time profiles of impact and reference means will follow parallel (but not necessarily identical) trajectories. In the **impact trend-by-time** design, trends in the relationship between resource levels and a continuous measure of perturbation magnitude are compared among repeated samples over time (Figure 6.2h). Although resource levels may vary over time due to natural variation, a significant relationship to the perturbation gradient is evidence of an impact. When there is no longer a significant trend on the gradient, recovery has occurred.

Both of these designs recognize that variation in environmental features or resources may occur over time and among sites. Assumptions of steady-state equilibria in time and space are therefore relaxed. Instead, one assumes a dynamic equilibrium between factors affecting the resource and the state of the resource. In the absence of the perturbation, resource levels would change in the same way over time in impact and reference areas or among sites on a perturbation gradient.

6.5   ON NATURAL VARIATION, PSEUDOREPLICATION, POWER AND SCALE

Pseudoreplication involves the use of inferential statistics to test for treatment effects when the treatments are not replicated or samples are not independent [31, 40]. It is a complicating factor in the design of all field studies in

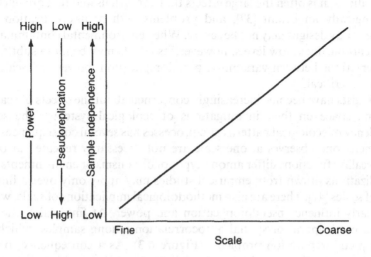

Figure 6.3 The relationship between the scale of sampling in time or space and independence among the samples, the likelihood of pseudoreplication and statistical power. See text for explanation.

environmental science, but it is of particular concern in assessing accidental environmental perturbations, where the treatment is by definition not replicated and where samples are constrained to fall within the contaminated area and thus may be neither temporally nor spatially independent [1]. The likelihood of pseudoreplication can be minimized by separating samples widely in time or space, as this will reduce intercorrelation among samples (Figure 6.3). On the other hand, if natural variation is great, then it will not require as much separation of samples in time or space to assure their independence. There is a paradox here: the variability that violates the assumptions of so many of the study designs may at the same time enhance the designs by reducing the likelihood of pseudoreplication.

Anyone assessing environmental impacts must also be concerned about statistical power, which relates to the likelihood of making Type II errors in statistical analyses and concluding that there is no effect of a perturbation when in fact there is one [24, 32, 41]. Power is a function of sample sizes, the effect size one wishes to document and the variability of a system. If natural variation is great, it may obscure the effects of a perturbation. One may then detect only large effects, or the sample size must be increased to retain power. However, because accidental perturbations are unplanned, sample size may often be limited. If natural variation is low, spatial and temporal covariation among samples may increase and samples must be more widely spaced to avoid blatant pseudoreplication. Consequently, one collects fewer samples and loses power (Figure 6.3); again, only large effects can be detected. With accidental perturbations, it is often the large effects that concern us and that are likely to be biologically important [37], and problems with pseudoreplication and power in study designs may not be severe. When environmental contamination occurs chronically at low levels, however, its effects may be more subtle and the interactions between variability, pseudoreplication, power and scale become more critical.

Ecologists have become increasingly concerned about the effects of scale in time and space on their investigations of ecological systems. The scale-dependency of ecological patterns and processes has several consequences: the phenomena one observes at one scale are not necessarily repeated at other scales, scaling functions differ among types of organisms or environments and generalizations drawn from empirical studies may apply only over a limited range of scales [42]. There are also methodological implications of scale, which particularly influence pseudoreplication and power. At fine scales, there is likely to be temporal or spatial autocorrelation among samples, which increases pseudoreplication problems (Figure 6.3). As a consequence, power may be low, either because the pseudoreplicated samples are not true samples or because, in an attempt to reduce pseudoreplication, fewer samples are taken on scales sufficient to ensure independence. These interrelated issues of scale, pseudoreplication and power are often neglected in designing studies of environmental impacts, but they are of fundamental importance.

6.6 CONCLUSIONS

It is obvious that natural variability complicates attempts to assess the impacts of environmental perturbations. Coping with this variability requires that one recognizes that natural systems generally do not adhere to the equilibrium conditions projected by theory or assumed in study designs. The study designs considered here, however, differ in their dependence on equilibrium assumptions. Unfortunately, the most straightforward (and most popular) designs are probably the least useful. They are founded on assumptions either that temporal variation is unimportant (baseline and time-series designs) or that there is little variation in environmental factors in space (impact–reference and gradient designs). Paired designs (matched pairs, pre/post pairs) reduce the effects of spatial variation in environmental factors, while more complex (and more demanding) designs (impact level-by-time, impact trend-by-time) allow one to assess the contributions of both spatial and temporal variation to the apparent impact. Inclusion of covariates in the analyses may also enhance one's ability to separate natural variability from perturbation effects. These latter two designs also are relatively robust to the effects of pseudoreplication [1]. Unfortunately, there are no simple ways to circumvent the complicating effects of natural variability in assessing environmental impacts.

Natural variability also affects the value and design of monitoring programs that may be implemented in anticipation of environmental impacts or to gauge recovery following a perturbation. Effective monitoring must distinguish between the effects of a perturbation and those of natural variation [26]. How long monitoring must be continued depends on the precision of estimates (true sampling error), the sample sizes that can be generated and the magnitude of resource changes likely to result from a perturbation in relation to the magnitude of natural variation. These, of course, are the elements of statistical power. Some evaluations of power in variable systems provide a sobering message to those expecting quick results from monitoring. Peterman [43], for example, calculated that it might require 30 years of data to recognize a 50% reduction in whale abundance using conventional estimation procedures. Krebs [44] has argued persuasively that the value of monitoring programmes may be enhanced considerably if experimental manipulations are part of the study design (but see also [45]). Although monitoring programmes may be most readily implemented when one wishes to assess the effects of planned perturbations, or determine recovery after a known impact, these may be the situations in which the effect size is most subtle and in which natural variation poses the greatest constraints. Monitoring may be more effective in detecting large changes in a system, such as those following an accidental perturbation. It seems to be one of those 'rules', however, that accidents rarely happen where one has a careful monitoring programme in place. Natural variability makes monitoring difficult to design, but without careful design monitoring may be politically attractive but ecologically useless [44].

Variability is the normal state of nature. The challenge of environmental impact or risk assessment is to determine how resources are affected by anthropogenic perturbations against this backdrop of variability. There are ways to cope with this variability, but they require careful attention to study design and the demands of statistical power. Investigators must consider the assumptions about variability contained in various study designs and especially avoid those that assume the system is in a steady-state equilibrium.

In view of all of these complications, the alternative of gauging environmental impacts using qualitative, subjective approaches may seem attractive. Human perturbations of the environment are invariably contentious and emotionally charged, however. Our responsibility as scientists is to provide the most objective and rigorous assessment of environmental effects possible, without which decision-makers have little but guesses, emotions and politics to guide them.

ACKNOWLEDGEMENTS

I thank Lorraine Maltby and Donald Baird for inviting me to participate in the SETAC–Europe, UK Branch conference at the University of Sheffield. This chapter is based on my presentation there. My thinking on this topic evolved through discussions with Keith Parker.

REFERENCES

1. Wiens, J. A. and Parker, K. R. (1995) Analyzing the effects of accidental environmental impacts: approaches and assumptions. *Ecological Applications*, **5**, 1069–83.
2. Egerton, F. N. (1973) Changing concepts of the balance of nature. *Quarterly Review of Biology*, **48**, 322–50.
3. Wiens, J. A. (1984) On understanding a non-equilibrium world: myth and reality in community patterns and processes, in *Ecological Communities: Conceptual Issues and the Evidence* (eds D. R. Strong, D. Simberloff, L. G. Abele and A. B. Thistle), Princeton University Press, Princeton, pp. 439–57.
4. DeAngelis, D. L. and Waterhouse, J. C. (1987) Equilibrium and nonequilibrium concepts in ecological models. *Ecological Monographs*, **57**, 1–21.
5. Chesson, P. L. and Case, T. J. (1986) Overview: nonequilibrium community theories: chance, variability, history, and coexistence, in *Community Ecology* (eds J. Diamond and T. J. Case), Harper and Row, New York, pp. 229–39.
6. Crawley, M. J. (1990) The population dynamics of plants. *Philosophical Transactions of the Royal Society of London B*, **330**, 125–40.
7. Shepherd, J. G. and Cushing, D. H. (1990) Regulation in fish populations: myth or mirage? *Philosophical Transactions of the Royal Society of London B*, **330**, 151–64.
8. Marchant, J. H., Hudson, R., Carter, S. P. and Whittington, P. (1990) *Population Trends in British Breeding Birds*, British Trust for Ornithology, Tring, Hertfordshire, UK.
9. Grieg-Smith, P. (1979) Pattern in vegetation. *Journal of Ecology*, **55**, 483–503.
10. Grieg-Smith, P. (1983) *Quantitative Plant Ecology*, 3rd edn, University of California Press, Berkeley.
11. Kolasa, J. and Pickett, S. T. A. (eds) (1991) *Ecological Heterogeneity*, Springer-Verlag, New York.
12. Wiens, J. A. (1995) Landscape mosaics and ecological theory, in *Mosaic Landscapes and Ecological Processes* (eds L. Hansson, L. Fahrig and G. Merriam), Chapman & Hall, London, pp. 1–26.
13. Gilpin, M. and Hanski, I. (1991) *Metapopulation Dynamics: Empirical and Theoretical Investigations*, Academic Press, London.
14. Blakers, M., Davies, S. J. J. F. and Reilly, P. N. (1984) *The Atlas of Australian Birds*, Melbourne University Press, Melbourne.
15. Cyrus, D. and Robson, N. (1980) *Bird Atlas of Natal*, University of Natal Press, Pietermaritzburg.
16. Gibbons, D. W., Reid, J. B. and Chapman, R. A. (1993) *The New Atlas of Breeding Birds in Britain and Ireland: 1988–1991*, T & AD Poyser, London.
17. Hyytiä, K., Koistinen, J. and Kellomäki, E. (1983) *Suomen Lintuatlas*, Lintutieto Oy, Helsinki.
18. Robbins, C. S., Bystrak, D. and Geissler, P. H. (1986) *The Breeding Bird Survey: Its First Fifteen Years, 1965–1979*, United States Department of the Interior, Fish and Wildlife Service, Resource Publication 157, Washington, DC.
19. Levin, S. A. (1992) The problem of pattern and scale in ecology. *Ecology*, **73**, 1943–67.
20. Cairns Jr, J. (1990) Lack of theoretical basis for predicting rate and pathways of recovery. *Environmental Management*, **14**, 517–26.
21. Hastings, A., Hom, C. L., Ellner, S., Turchin, P. and Godfray, H. C. J. (1993) Chaos in ecology: is mother nature a strange attractor? *Annual Review of Ecology and Systematics*, **24**, 1–33.

22. Hastings, A. and Higgins, K. (1994) Persistence of transients in spatially structured ecological models. *Science*, **263**, 1133–6.
23. Link, W. A. and Nichols, J. D. (1994) On the importance of sampling variance to investigations of temporal variation in animal population size. *Oikos*, **69**, 539–44.
24. Osenberg, C. W., Schmitt, R. J., Holbrook, S. J., Anu-Saba, K. E. and Flegal, A. R. (1994) Detection of environmental impacts: natural variability, effect size, and power analysis. *Ecological Applications*, **4**, 16–30.
25. Schroeter, S. C., Dixon, J. D., Kastendiek, J. and Smith, R. O. (1993) Detecting the ecological effects of environmental impacts: a case study of kelp forest invertebrates. *Ecological Applications*, **3**, 331–50.
26. Skalski, J. R. and McKenzie, D. H. (1982) A design for aquatic monitoring programs. *Journal of Environmental Management*, **14**, 237–51.
27. *Exxon Valdez* Trustee Council (1993) Draft Restoration Plan. *Exxon Valdez* Trustee Council, Anchorage, AK.
28. Wiens, J. A. (1995) Recovery of seabirds following the *Exxon Valdez* oil spill: an overview, in *The* Exxon Valdez *Oil Spill: Environmental Impact and Recovery Assessment* (eds P. G. Wells, J. N. Butler and J. S. Hughes), Special Technical Publication 1219, American Society for Testing and Materials, Philadelphia, PA, pp. 854–93.
29. Dutilleul, P. (1993) Spatial heterogeneity and the design of ecological field experiments. *Ecology*, **74**, 1646–58.
30. Eberhardt, L. L. and Thomas, J. M. (1991) Designing environmental field studies. *Ecological Monographs*, **61**, 53–73.
31. Hurlbert, S. H. (1984) Pseudoreplication and the design of ecological field experiments. *Ecological Monographs*, **54**, 187–211.
32. Morrisey, D. J. (1993) Environmental impact assessment – a review of its aims and recent developments. *Marine Pollution Bulletin*, **26**, 540–5.
33. Shrader-Frechette, K. S. and McCoy, E. D. (1993) *Methods in Ecology: Strategies for Conservation*, Cambridge University Press, Cambridge.
34. Skalski, J. R. and Robson, D. S. (1992) *Techniques for Wildlife Investigations*, Academic Press, San Diego, CA.
35. Stewart-Oaten, A., Bence, J. R. and Osenberg, C. W. (1992) Assessing effects of unreplicated perturbations: no simple solutions. *Ecology*, **73**, 1396–404.
36. Underwood, A. J. (1993) The mechanics of spatially replicated sampling programmes to detect environmental impacts in a variable world. *Australian Journal of Ecology*, **18**, 99–116.
37. Underwood, A. J. (1994) On beyond BACI: sampling designs that might reliably detect environmental disturbances. *Ecological Applications*, **4**, 3–15.
38. Day, R. H., Murphy, S. M., Wiens, J. A., Hayward, G. D., Harner, E. J. and Smith, L. N. (1995) Use of oil-affected habitats by birds after the *Exxon Valdez* oil spill, in *The* Exxon Valdez *Oil Spill: Environmental Impact and Recovery Assessment* (eds P. G. Wells, J. N. Butler and J. S. Hughes), Special Technical Publication 1219, American Society for Testing and Materials, Philadelphia, PA, pp. 726–61.
39. Magnuson, J. J., Benson, B. J. and Kratz, T. K. (1990) Temporal coherence in the limnology of a suite of lakes in Wisconsin, U.S.A. *Freshwater Biology*, **23**, 145–59.
40. Stewart-Oaten, A., Murdoch, W. W. and Parker, K. R. (1986) Environmental impact assessment: 'pseudoreplication' in time? *Ecology*, **67**, 929–40.
41. Fairweather, P. G. (1993) Links between ecology and ecophilosophy, ethics and the requirements of environmental management. *Australian Journal of Ecology*, **18**, 3–19.

42. Wiens, J. A. (1989) Spatial scaling in ecology. *Functional Ecology*, **3**, 385–97.
43. Peterman, R. M. (1989) Statistical power analysis can improve fisheries research and management. *Canadian Journal of Fisheries and Aquatic Sciences*, **47**, 2–15.
44. Krebs, C. J. (1991) The experimental paradigm and long-term population studies. *Ibis*, **133** (Suppl.), 3–8.
45. Greenwood, J. J. D. and Baillie, S. R. (1991) Effects of density-dependence and weather on population changes of English passerines using a non-experimental paradigm. *Ibis*, **133** (Suppl.), 121–33.

7 Environmental stress and the distribution of traits within populations

VALERY E. FORBES AND MICHAEL H. DEPLEDGE

Here we explore the nature of the relationship between environmental stress and the distribution of relevant population attributes. Additionally, we examine the meaning of the term 'sensitivity' as used by ecologists versus toxicologists. In toxicology, the sensitivity of a population to an environmental stress is described by the location (e.g. mean or median) of the trait distribution with reference to an environmental gradient, whereas in ecology sensitivity is defined by the shape (i.e. variance) of the distribution across an environmental gradient. Recognition of this discrepancy has important implications with regard to the interpretation of population tolerance patterns. At various levels of biological organization, changes in the average or median value of biological structure or function in response to stress may be accompanied by changes in variance. We review selected studies from the fields of molecular genetics, physiology, whole organism energetics, population biology and ecosystem ecology, suggesting that changes in the variance of population attributes may indicate a general symptom of environmental stress. Stress-caused increases in biological variability will decrease our ability to detect effects of stress, particularly when the focus is on population means. Analysis of distributional shape changes of relevant population attributes may offer a sensitive approach for assessing and predicting the impact of chemical toxicants.

ECOtoxicology: Ecological Dimensions. Edited by D.J. Baird, L. Maltby, P.W. Greig-Smith and P.E.T. Douben. Published in 1996 by Chapman & Hall, London. ISBN hardback 0 412 754970 3 and paperback 0 412 75490 8.

7.1   ARE RESPONSES TO ENVIRONMENTAL STRESS GENERAL?

To what extent is it possible to generalize responses to chemical stress within and among species? It has become painfully obvious that response parameters such as the LC_{50} (median lethal concentration), EC_{50} (median effective concentration), LOEC (lowest observed effect concentration) and NOEC (no observed effect concentration) can vary by up to several orders of magnitude in ways that are often difficult to predict *a priori*. Yet, at some scales and for some purposes making generalizations about the way organisms respond to chemical stresses is desirable, and even essential. In this chapter we examine the relationship between environmental stress and the distribution of population traits. Specifically, we postulate that the variance of relevant population attributes influences and is influenced by the effective stress experienced by a population.

7.2   KEY DEFINITIONS

A few definitions are necessary before we begin. We use **environmental stress** to refer to an external factor or constraint that limits the activities of organisms. Implicit in our usage of the term stress are the assumptions that the activities of organisms: (1) are related to their potential for contributing to the next generation (i.e. their fitness); (2) are influenced by their molecular and cellular makeup, and by their physiological condition; and (3) contribute to the structures and processes characteristic of populations, communities and ecosystems. Thus, responses to environmental stress are exhibited at all levels of biological organization over a wide range of spatial and temporal scales. Biological systems can become stressed by fluctuations in natural variables, such as temperature, oxygen or food level, and by human-imposed alterations in their environment, such as chemical toxicants, noise or physical habitat alteration.

   Much of the following discussion will deal with the variance of traits within populations of individuals of single species (i.e. biological populations) and the relationship between population variance and environmental stress. However, we extend discussion beyond biological populations, and note that our use of the term **population** is in its statistical sense – 'the totality of individual observations about which inferences are to be made, existing anywhere in the world or at least within a definitely specified sampling area limited in space and time' [1, p. 9]. Used in this way, we can sample from a population of DNA contents in the cells of all white rats in the world, a population of heart rates in crabs from a Danish beach, a population of photosynthetic rates in all temperate forest ecosystems. Such a definition facilitates comparison of sample distributions across levels of biological organization. **Trait** is simply used to mean the population attribute or variable under study (e.g. heart rate, body size, abundance, photosynthetic rate, etc.).

In the fields of ecology and evolution, variability in the response of organisms to environmental conditions has long provided a focus for study and underlies basic ecological concepts such as the niche. Natural selection requires the existence of variability among individuals or genotypes, and therefore examination of the distribution of traits in populations, and changes of such in space and time have long been active areas of research. Lynch and Gabriel [2] defined a **tolerance** curve as the response of a genotype's total fitness over an environmental gradient. The tolerance curve is defined by the genotype's environmental optimum and a measure of the equitability of fitness over the environmental gradient. Both the shape (variance) and location (mean or median) of the tolerance curve may evolve in response to selection.

7.3 POPULATION TOLERANCE DISTRIBUTIONS

As all organisms are composed of similar constituents, we can expect a certain degree of generality in response to chemical toxicants and other environmental stresses. For example, as the dose of a drug or toxic chemical applied to a group of organisms is increased, the strength of response exhibited by the population (either in proportion of individuals responding or size of the average response) is typically observed to increase monotonically [3]. This relationship, in its familiar form as the dose–response relationship, provides the foundation for the study of pharmacology and toxicology. In ecotoxicology, the dose (or concentration)–response relationship provides the theoretical basis for the design and interpretation of virtually all research, assessment and management activities. This is true whether the emphasis is on organism components, single species, communities or ecosystems.

The quantal (for all-or-none type data) concentration–response relationship is derived from the idea that individual organisms in a population differ in their ability to tolerate exposure to imposed stimuli. The frequency distribution of tolerances across a chemical gradient typically approximates a log normal distribution (Figure 7.1a). The ideal tolerance distribution, like all normal distributions, is completely characterized by only two parameters, a mean (or median) and a standard deviation. The mean and standard deviation are most effectively estimated by transforming the tolerance curve to a linear form, e.g. by the probit transformation (Figure 7.1b).

Trevan [3] first suggested using the dose–response relationship to estimate the relative potency of chemicals to test populations. He noted that variability among individuals in their response to a chemical led to a characteristic S-shaped dose–response curve (when response is expressed as a cumulative frequency). Because variability among individuals was least at about the 50% level of response, Trevan recommended characterizing a chemical by its median effective dose – the dose that on average produces a response in 50% of the test population – and to this day the median response of a population to a chemical stimulus remains the most widely used response measure in

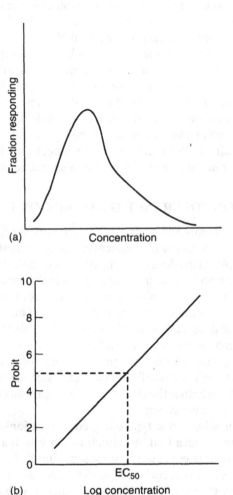

Figure 7.1 A typical concentration–response curve for quantal data. (a) The population tolerance distribution shows the fraction of the population just responding at each concentration. (b) Data from (a) linearized by a probit transformation. The EC_{50} is the chemical concentration at which half the population responds. The slope of the probit curve is equal to the reciprocal of the standard deviation of the population tolerance distribution.

ecotoxicological studies. The dose (or concentration)–response relationship is a very general phenomenon in that it can be fitted to a wide variety of biological responses to a multitude of chemicals. However, its general nature is of limited usefulness in that both biological and environmental factors substantially influence its shape and location.

7.4 SENSITIVITY IN TOXICOLOGY

The importance of studying intrapopulation variability in response to pollutant exposure was highlighted by Depledge [4]. However, toxicologists and ecologists view such variability from different perspectives and consequently differ in how they define a population's 'sensitivity' to environmental stress.

Finney [5] adopted the definition of sensitivity used by pharmacologists, namely, the minimal dose that produces an observable response (i.e. sensitivity is defined by the location of the tolerance distribution). In contrast, he used the variance of the tolerance distribution to measure the 'responsiveness' or uniformity of the population. Figure 7.2 shows a linear transformation of the tolerances of populations A–C to a chemical gradient. The y-axis represents a measure of growth relative to an unexposed control group. The vertical lines drawn to the x-axis represent the concentration at which growth of each population is inhibited by 50% (EC_{50}). Defining each population's sensitivity by its median response leads to the conclusion that populations A and B are equally sensitive to the chemical under consideration, whereas population C, with its lower EC_{50}, is more sensitive. Population A, with its steeper slope, would have smaller confidence limits around its calculated EC_{50}, relative to populations B and C, and would therefore be interpreted as being more uniform in its response (or more responsive in Finney's terminology).

7.5 SENSITIVITY IN ECOLOGY

In quantitative genetic studies, environmental sensitivity of a genotype or population is calculated as the variance in performance across an environmental

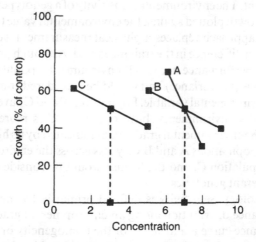

Figure 7.2 Response of three populations, A, B and C, to a chemical gradient plotted on a linear scale. The response is growth inhibition relative to an unexposed control population. The vertical lines drawn to the abscissa are the EC_{50} values for each population (the concentration at which growth is reduced by 50%).

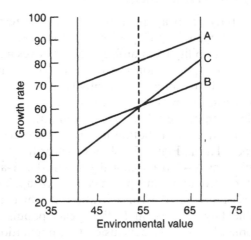

Figure 7.3 The performance of three genotypes, A, B and C, along an environmental gradient. The measure of performance is growth (in arbitrary units). The environmental value is calculated as the average performance of all of the genotypes in each environment. The dashed vertical line marks the mean performance in the different environments. See text for details.

gradient [6]. Figure 7.3 shows growth rates of three genotypes along an environmental gradient. The sensitivities of the different genotypes are quantified by giving each environment a value equal to the mean growth rate of all genotypes in that environment. The environmental sensitivity of a genotype is then the slope of the line of its growth plotted against the environmental value (see Falconer [6] for details). This approach replaces a physical measurement, such as chemical concentration, by a difference in the trait measured. Although population A has a higher average performance across all environments, populations A and B have the same slope (i.e. variance) and would be considered equally sensitive to changes in this environmental variable. Populations B and C have the same mean performance across environments, but population C is more sensitive than population B to the environmental gradient, as indicated by its higher slope. The performances of populations A and B vary less across the environmental gradient than does population C, and the former would be considered to represent more broadly tolerant genotypes.

Thus, in toxicology, sensitivity is defined primarily by the median of the population tolerance distribution along an environmental gradient. The variance of the tolerance curve is a measure of the homogeneity or responsiveness of the test population. In ecology, sensitivity is defined by the variance of the tolerance distribution, irrespective of its median, with respect to an environmental gradient. A population or genotype with a narrower tolerance distribution (a higher slope) is more sensitive to the effect of the environmental stress whereas a wide tolerance curve is interpreted as representing a broadly adapted or general purpose genotype [7].

The distinction is important for several reasons. First, since ecotoxicology ideally integrates the fields of ecology and toxicology, it is essential that our terminology be consistent across disciplines. Second, because both the location and shape of the tolerance distribution are essential components of the response of biological systems to environmental stress, it is important that these parameters be quantified and interpreted correctly [8, 9]. Third, illuminating the sources of intrapopulation variability can aid in developing a mechanistic understanding of observed responses to particular stresses. Soares *et al.* [10] used the quantitative genetics approach to examine interclonal variation in the tolerance of *Daphnia magna* to chemical toxicants. They were able to partition the variability in stress tolerance among clones into genetic, environmental and interaction effects, and could thereby gain insight into the causes of variability in *D. magna*'s response to chemical stress.

7.5.1 CAUSES OF VARIABILITY WITHIN AND AMONG POPULATIONS

In genetically heterogeneous populations, genetic influences are often assumed to be responsible for a significant fraction of the intrapopulation variability in response to chemical stresses. In toxicology it is typically assumed that reducing intrapopulation genetic variability will necessarily result in a more homogenous response, i.e. a steeper concentration–response curve. Early studies indicated that inbred animals, such as Wistar rats, yielded steeper dose–response curves than did genetically mixed animals [11]. Although still a commonly held assumption, there are both empirical and theoretical reasons for questioning its validity (cited in [5], p. 405; [13–17]). For example, inbreeding involves the loss of heterozygosity [12] and inbred lines are often more phenotypically variable than their outbred ancestors [13].

Alternatively, asexual taxa appear to be more heterozygous than their sexual relatives either because their origin involved polyploidy [14] or because the reproduction of an intact genome is believed to involve the accumulation of mutations [15]. The idea that the increased genetic heterozygosity of asexual organisms is reflected in a broadly adapted or general purpose genotype is supported by the observation that asexual organisms often occupy a wider range of environments than their sexual relatives and are more frequent occupants of marginal or peripheral habitats [7, 15, 16 and references therein]. However, direct evidence for the general purpose genotype hypothesis remains equivocal. Tests of this hypothesis most often involve comparison of the coefficients of variation (CV) of selected traits between sexual and asexual populations. A lower CV is interpreted to indicate less environmental sensitivity (i.e. a more general purpose response, e.g. [17]).

7.6   VARIABILITY IN ASEXUAL VERSUS SEXUAL POPULATIONS

Browne *et al.* [18] measured 12 life history traits in 12 populations of the brine shrimp, *Artemia*, under controlled laboratory conditions (Table 7.1). Five of the populations were made up of single clones from obligately parthenogenetic populations and seven were made up of genetically mixed individuals from sexually reproducing populations. It should be emphasized that the traits were not measured across an environmental gradient but in a uniform laboratory setting. They estimated the fraction of the variance in life history traits due to environmental variation by dividing the average CV of the asexual populations by the average CV of the sexual populations (i.e. by assuming that all of the variance in the asexual populations is due to uncontrolled environmental influences). The percentage of the variance attributable to genetic factors was calculated as 100% minus the percent environmental variation. We compared the average CVs between sexual and asexual groups (using data presented in their table 4). The first point worth noting is that different life history traits differed substantially in variability. Overall, the least variable trait (lowest CV in eight of ten populations) was the female pre-reproductive period. In contrast, the female post-reproductive period was one of the most variable traits with CVs exceeding 100% in ten of the 12 populations. The percentages of offspring encysted and cysts hatched were also highly variable. Of the 11 traits that could be compared between sexual and asexual populations (male life span deleted), five showed no significant difference in variance between sexual and

Table 7.1   Mean coefficients of variation (CVs), adjusted for sample size, for life history characteristics in sexual and asexual populations of the brine shrimp, *Artemia*. The means are based on seven sexual populations and five asexual populations

Trait measured	Mean sexual CV (%)	Mean asexual CV (%)
1 Offspring per brood	43[a]	24
2 Broods per female	56	43
3 Offspring per day per female	62[a]	25
4 Days between broods	36[a]	16
5 Offspring encysted (%)	64	142[b]
6 Cysts hatched (%)	93	117
7 Total offspring per female	83[a]	48
8 Female prereproductive period (d)	16	7
9 Female reproductive period (d)	69	47
10 Female postreproductive period (d)	115	151[b]
11 Total female life span	33	32
12 Total male life span	41	–

[a]CV significantly greater for sexuals.
[b]CV significantly greater for asexuals, remaining CVs did not differ between groups. Data from [18].

asexual populations (traits 2, 6, 8, 9, 11). Two traits (5 and 10) were more variable in asexuals, whereas four traits (1, 3, 4, 7) were more variable in sexuals.

These results are important because they demonstrate that even under supposedly optimal and uniform laboratory conditions, single genotypes can exhibit substantial amounts of variability in fitness-related traits. For some traits measured by Browne *et al.* [18], not only was 100% of the variability attributed to uncontrolled environmental influences, but for two traits the variances were actually higher in populations of single clones than in populations of genetically mixed individuals, implying a greater sensitivity of the asexual genotypes to uncontrolled microhabitat heterogeneity with regard to these traits. Intraclonal variability within a single environment has also been attributed to random noise in developmental pathways, maternal effects or physiological acclimation [2].

7.7 STRESS-CAUSED CHANGES IN VARIABILITY

As discussed above, the variance of relevant population traits is used in toxicology and ecotoxicology to estimate the population's uniformity or responsiveness to an imposed stress. A less recognized, yet potentially important, issue is that the variance of relevant traits can itself change in response to environmental stress. There are several ways that stress can influence the variance of phenotypic traits within populations [19] – by causing mutation, by inducing the expression of genes already present and by increasing recombination rates.

Evidence from the early 1900s suggested that the incidence of recombination increases as the environment becomes more stressful (cited in [19]). Borodin [20] suggested that hormones released in response to stress induce an inhibition of replication and repair synthesis of DNA which may disrupt synapsis and influence recombination. Walbot and Cullis [21] argued that environmental conditions can induce mutational changes in plants, many of which may be translated into phenotypic variation. Bradshaw and Hardwick [22] argued that when stress varies temporally or spatially over scales experienced by individual organisms, phenotypic plasticity, rather than selection for purely tolerant genotypes, is a more likely evolutionary pathway.

Development and application of techniques such as PCR and flow cytometry by molecular biologists have allowed a more detailed investigation of the effects of stress on mutation rates. Using flow cytometric techniques, Bickham [23] and Otto *et al.* [24] showed that mutagenic chemicals and ionizing radiation increase variability of nuclear or chromosomal DNA content in a positive dose–response relationship, both *in vivo* and *in vitro*.

The idea that *de novo* variation due to recombination and mutation in the broadest sense increases under stress is supported by some recent evidence (reviewed by Parsons [25]). He writes 'Hence at stressful moments in evolutionary history when there is a premium on major adaptive shifts, variability of all types may be increased, and this could trigger genomic reorganizations in response to rapidly changing environments.'

7.8 EVIDENCE FOR INCREASED VARIABILITY IN RESPONSE TO STRESS?

Stress has been shown to increase the variability of biological attributes at many levels of organization. Work in our laboratories over the last several years has focused on the levels and sources of intrapopulation variability in physiological traits of benthic invertebrates exposed to heavy metals. Forbes *et al.* [26] found that in both unexposed control treatments and cadmium-exposed treatments, the coefficients of variation of growth were at least as high within single clones of the parthenogenetic gastropod, *Potamopyrgus antipodarum*, as within populations composed of genetically mixed snails. Exposure to cadmium increased the variability in growth rate in all populations and altered the ranking of variabilities among populations (Table 7.2).

Bjerregaard [27] examined the effect of physiological condition on cadmium accumulation and binding in tissues of the shore crab, *Carcinus maenas*. Although he showed that the means of ten physiological traits were generally similar between control and exposed groups (only six of the 45 exposed group means shown here were found to be significantly different by Bjerregaard [27]), there was an indication that the variance in physiological condition was greater in exposed than in control crabs. From Bjerregaard's data we have calculated CVs for each of the physiological traits as a function of cadmium concentration (Table 7.3). We calculated the sign of the difference in CVs between the control group and each exposure group for each of the ten traits shown in Table 7.3. We assumed that if there is no

Table 7.2 Mean growth (mm shell length) of four species of *Hydrobia* and two populations of *Potamopyrgus antipodarum* during three weeks in control and cadmium (200 mg dm^{-3}) treatments. Coefficients of variation have been corrected for sample size [1, p. 58]. Sample size for the *Hydrobia* species is between 11 and 16. Data for *P. antipodarum* represent within-clone averages calculated for 6–8 clones with 3–4 individuals per clone. From [26]

Species	Control growth (SD)	CV	Cadmium growth (SD)	CV
H. ventrosa	1.41 (0.37)	0.27	1.01 (0.34)	0.34
H. truncata	1.05 (0.29)	0.28	0.58 (0.17)	0.31
H. ulvae	2.13 (0.34)	0.16	0.43 (0.27)	0.65
H. neglecta	1.67 (0.29)	0.18	0.46 (0.11)	0.24
P. antipodarum				
freshwater	0.54 (0.11)	0.21	0.11 (0.03)	0.25
brackish	0.24 (0.09)	0.32	0.06 (0.04)	0.70

Table 7.3 Coefficients of variation (CVs), corrected for sample size, for physiological traits measured in groups of *Carcinus maenas* exposed to different concentrations of cadmium. Data are from table 2 in Bjerregaard [27]. Traits 1 and 2 apply to the haemolymph only, whereas the remaining traits were measured in the hepatopancreas only

Cadmium (mg Cd dm^{-3})	0	0.25	0.5	0.75	1.0	1.5
Trait						
1 Vol. (% body wt)	16.3	16.4	12.3	25.3	15.4	10.6
2 Protein (mg cm^{-3})	15.2	20.3	15.1	15.5	58.2	18.1
3 Zn (mg kg^{-1})	36.5	12.0	14.4	10.9	15.1	23.8
4 Ca (mol kg^{-1})	9.2	12.0	34.7	33.6	43.9	47.8
5 Mg (mmol kg^{-1})	7.6	5.1	36.2	24.4	33.2	38.6
6 Soluble protein (mg g^{-1})	21.8	11.5	25.2	8.7	21.3	13.9
7 Wet hepatopan. of body wet weight (%)	7.8	19.5	19.5	16.2	22.3	19.8
8 Dry hepatopan. of body wet weight (%)	10.6	8.0	32.1	8.9	31.2	29.0
9 Wet wt:dry wt ratio	8.0	11.3	13.4	13.3	13.9	15.5
10 Cu (mg kg^{-1})[a]	99	649	321	99	140	175

[a]Values shown are the range [1, p. 48] as this trait was not normally distributed.

effect of cadmium on physiological trait variability, there is an equal probability of obtaining either a positive or a negative difference between the CV of the control group and the CV of cadmium exposed groups. Thirty-four out of a total of 50 differences, or 68%, of the CVs were higher in the exposed than in the control groups. The probability of such an outcome is less than 5%, indicating that it is unlikely that the control crabs are as physiologically variable as the exposed crabs. The CVs for some of the traits (e.g. calcium concentration in the hepatopancreas and wet weight:dry weight ratios) exhibit a monotonically increasing concentration–response relationship over the range of concentrations measured. For other traits such a trend is not apparent, but these data suggest that a more systematic analysis of the effects of environmental stress on intrapopulation physiological variability would be worthwhile.

At higher levels, stress is manifested by reductions in the stability and diversity of ecosystems [28]. Both spatial and temporal fluctuations in population size appear to increase in response to stress. Warwick and Clarke [29] found that the variability of species abundances among replicate samples was greater in impacted sites relative to control sites. They suggested that the within-site variability in community structure may represent an identifiable symptom of disturbed systems. Rapport *et al.* [30] discussed the observation that population fluctuations tend to be greater in stressed ecosystems. Such instability in abundance can result from a destabilization of population buffering mechanisms or may be partly a consequence of a shift to shorter-lived, opportunistic species in stressed environments. Breakdown of ecosystem feedback mechanisms, which dampen fluctuations, can be a symptom of ecosystem stress [31], evidenced by large fluctuations in ecosystem attributes such as biomass, productivity, energy transfer and nutrient cycling.

7.9 PREDICTING VARIANCE CHANGES IN RESPONSE TO STRESS

Whether the variance of key population traits increases or decreases in response to stress is likely to depend on the trait measured as well as on the degree and duration of stress. Also, populations differ in inertia, resilience and stability, all of which will influence how abundance varies in time and space as well as influencing the population's response to stress [32]. Although there is evidence that a certain level of stress appears to increase variability, an increase in the level of that stress can cause death, thereby eliminating variation. Thus, Parsons [25] argued that the margin between high variability and the potential for extinction is likely to be extremely narrow.

The nature of environmental tolerance curves implies that fluctuating environmental conditions encourage the evolution of generalism, i.e. they tend to decrease a genotype's sensitivity (in the ecological sense) to environmental fluctuations. Apparently temporal variation within generations plays a more important role than that between generations, particularly when the spatial component of environmental variation is high [2].

It is likely that changes in the variance of the population tolerance distribution vary with time during periods of exposure to stress. Depending on when the stressed population is sampled, the median of the tolerance distribution may or may not have shifted and the variance in the tolerance distribution may be greater or less than in an unstressed control population. For example, the evolution of resistance to insecticides often appears to arise via the action of a single resistance gene and typically exhibits the following pattern (Figure 7.4). In the early stages of selection, when the frequency of resistance genes is low (c. 10%) the LC_{50} of the exposed population may be similar to an unselected control population, but the slope of the concentration–response curve decreases. The lowered slope is indicative of a heterogeneous (i.e. partially resistant) population and has been proposed as the best indicator of incipient resistance [9, 33]. As the frequency of the resistance gene increases to fixation in the selected population, the slope of the concentration–response curve is expected to increase (possibly to higher values than in the population prior to selection) whereas the location of the curve would be shifted to the right (i.e. to a higher LC_{50}). Changing tolerance patterns such as these were demonstrated to occur in Danish populations of the housefly, *Musca domestica*, exposed to the organophosphate insecticide tetrachlorvinphos (cited in [33]).

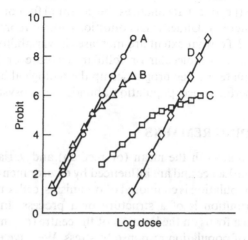

Figure 7.4 Simulated response of an hypothetical population evolving resistance to a toxic chemical. Curve A shows the expected concentration–response curve in an unselected control population. Curves B, C and D depict successive changes in the tolerance distribution as resistance evolves. Patterns such as this one have been described for insect populations evolving resistance to insecticide exposure [9, 33].

7.10 DETECTING EFFECTS OF STRESS ON VARIABLE SYSTEMS

To the extent that stress increases the variance in population traits, it impedes our ability to detect the effects of stress on biological systems. Both sample size and within-sample variability limit the power of statistical tests [9]. To the degree that stress increases sample variability it decreases our ability to detect effects that are there. Either our tests will be less powerful or we will be forced to increase sample size so that statistical power can be maintained. If variances differ between control and perturbed systems, the use of alternative (and frequently less powerful) statistical test methods may be necessary.

7.11 PROMISING DIRECTIONS FOR FUTURE INQUIRY

Although we are well aware that populations can exhibit widely differing tolerance distributions, very few general principles have been proposed to guide the prediction of tolerance differences *a priori*. Temporal or spatial changes in the shape of population tolerance distributions can reveal whether natural biological systems have or are likely to evolve tolerance to environmental stresses.

The recognition that environmental stress is intimately related to biological variability opens several avenues for further study. Under what conditions does a change in variability in response to environmental stress represent an adaptive response (i.e. one that enhances the potential for continued existence of the system of interest)? Under what conditions does it represent a symptom of system collapse? To what extent are increases in variability at lower levels of organization (e.g. in molecular or cellular structure or function) either amplified or attenuated as one progresses up the biological hierarchy (e.g. to whole organism performance, population abundance, ecosystem process)?

7.12   CONCLUDING REMARKS

We have argued that both the mean (or median) and variance of relevant population traits influence and are influenced by environmental stress. This is true whether the population is composed of organisms, cells or ecosystems and whether the distribution is of a structure or a process. In toxicology the tendency has been a focus on the location of the centre (mean or median) as a relevant measure of population response to stress. We have emphasized that the variance of the population distribution is also essential, particularly as it determines the potential for the evolution of tolerance to environmental stresses. There are indications at many different levels of biological organization that stress can alter biological variability. The direction and magnitude of the alteration are likely to depend on the degree and timescale of the stress applied. Thus, explicit analysis of changes in the variance of relevant population attributes may offer a sensitive approach for assessing and predicting the impact of chemical toxicants.

ACKNOWLEDGEMENTS

We thank Poul Bjerregaard, Tom Forbes and two anonymous reviewers for critical review of the manuscript.

REFERENCES

1. Sokal, R.R. and Rohlf, F.J. (1995) *Biometry: The Principles and Practice of Statistics in Biological Research*, 3rd edn, W.H. Freeman and Co., San Francisco.
2. Lynch, M. and Gabriel, W. (1987) Environmental tolerance. *The American Naturalist*, **129**, 283–303.
3. Trevan, J.W. (1927) The error of determination of toxicity. *Proceedings of the Royal Society, Series B*, **101**, 483–514.
4. Depledge, M.H. (1990) New approaches in ecotoxicology: can inter-individual physiological variability be used as a tool to investigate pollution effects? *Ambio*, **19**, 251–2.
5. Finney, D.J. (1978) *Statistical Method in Biological Assay*, 3rd edn, Charles Griffen and Co., London.
6. Falconer, D.S. (1990) Selection in different environments: effects on environmental sensitivity (reaction norm) and on mean performance. *Genetic Research, Cambridge*, **56**, 57–70.

7. Lynch, M. (1984) Destabilizing hybridization, general-purpose genotypes and geographic parthenogenesis. *Quarterly Review of Biology*, **59**, 257–90.
8. Forbes, T.L. (1993) The design and analysis of concentration-response experiments, in *Handbook of Ecotoxicology* (ed. P. Calow), Blackwell Scientific, Oxford, pp. 438–60.
9. Forbes, V.E. and Forbes, T.L. (1994) *Ecotoxicology in Theory and Practice*, Chapman & Hall, London.
10. Soares, A.M.V.M., Baird, D.J. and Calow, P. (1992) Interclonal variation in the performance of *Daphnia magna* Straus in chronic bioassays. *Environmental Toxicology & Chemistry*, **11**, 1477–83.
11. Gaddum, J.H. (1933) Reports on biological standards. III. Methods of biological assay depending on quantal response. *Medical Research Council, Special Report Series*, **183**.
12. Falconer, D.S. (1989) *Introduction to Quantitative Genetics*, 3rd. edn, Longman Scientific & Technical, New York.
13. Maynard Smith, J. (1989) *Evolutionary Genetics*, Oxford University Press, Oxford.
14. Stebbins Jr, G.L. (1950) *Variation and Evolution in Plants*, Columbia University Press, New York.
15. Bell, G. (1982) *The Masterpiece of Nature: The Evolution and Genetics of Sexuality*, University of California Press, Berkeley.
16. Bierzychudek, P. (1989) Environmental sensitivity of sexual and apomictic *Antennaria*: Do apomicts have general-purpose genotypes? *Evolution*, **43**, 1456–66.
17. Weider, L.J. (1993) A test of the 'general-purpose' genotype hypothesis: differential tolerance to thermal and salinity stress among *Daphnia* clones. *Evolution*, **47**, 965–9.
18. Browne, R.A., Sallee, S.E., Grosch, D.S., Segreti, W.O. and Purser, S.M. (1984) Partitioning genetic and environmental components of reproduction and lifespan in *Artemia. Ecology*, **65**, 949–60.
19. Hoffmann, A.A. and Parsons, P.A. (1991) *Evolutionary Genetics and Environmental Stress*, Oxford University Press, Oxford.
20. Borodin, P.M. (1987) Stress and genetic variability. *Genetika*, **23**, 1003–10.
21. Walbot, V. and Cullis, C.A. (1985) Rapid genomic change in higher plants. *Annual Review Plant Physiology*, **36**, 367–96.
22. Bradshaw, A.D. and Hardwick, K. (1989) Evolution and stress – genotypic and phenotypic components. *Biological Journal Linnean Society*, **37**, 137–55.
23. Bickham, J.W. (1990) Flow cytometry as a technique to monitor the effects of environmental genotoxins on wildlife populations, in *In Situ Evaluations of Biological Hazards of Environmental Pollutants* (eds S.S. Sandhu *et al.*), Plenum Press, New York, pp. 97–108.
24. Otto, F.J., Oldiges, H., Göhde, W. and Jain, V.K. (1981) Flow cytometric measurement of nuclear DNA content variations as a potential *in vivo* mutagenicity test. *Cytometry*, **2**, 189–91.
25. Parsons, P.A. (1989) Environmental stresses and conservation of natural populations. *Annual Review of Ecology & Systematics*, **20**, 29–49.
26. Forbes, V.E., Møller, V. and Depledge, M.H. (1995) Intrapopulation variability in sublethal response to heavy metal stress in sexual and asexual gastropod populations. *Functional Ecology*, **9**, 477–84.
27. Bjerregaard, P. (1990) Influence of physiological condition on cadmium transport from haemolymph to hepatopancreas in *Carcinus maenas. Marine Biology*, **106**, 199–209.

28. Regier, H.A. and Cowell, E.B. (1972) Application of ecosystem theory: succession, diversity, stability, stress and conservation. *Biological Conservation*, **4**, 83–8.
29. Warwick, R.M. and Clarke, K.R. (1993) Increased variability as a symptom of stress in marine communities. *Journal of Experimental and Marine Biology and Ecology*, **172**, 215–26.
30. Rapport, D.J., Regier, H.A. and Hutchinson, T.C. (1985) Ecosystem behavior under stress. *American Naturalist*, **125**, 617–40.
31. Schaeffer, D.J., Herricks, E.E. and Kerster, H.W. (1988) Ecosystem Health: I. Measuring ecosystem health. *Environmental Management*, **12**, 445–55.
32. Underwood, A.J. (1989) The analysis of stress in natural populations. *Biological Journal of the Linnean Society*, **37**, 51–78.
33. Wood, R.J. (1981) Insecticide resistance: genes and mechanisms, in *Genetic Consequences of Man Made Change* (eds J.A. Bishop and L.M. Cook), Academic Press, London, pp. 53–96.